1

Space Power Systems
Fourth edition

Copyright 2018 by Jeff Mitchell

Published by R. S. Publications
4112 Eagle Feather Drive
Orlando, Fl 32829

Cover art Courtesy NASA.

ISBN-978-0-9726926-3-2

Table of Contents

Acknowledgments

The author would like to thank all his students who, over the years, told him to turn his notes into a book. To the late Doctor John Clark, former head of the Goddard Space Flight Center and mentor. Also to his fellow professors, who teach at the Florida Institute of Technology and educate the graduate students in Orbital Mechanics, Rocket Propulsion, Space Communication, and other tough subjects. To Lieutenant Commander Scott B. Schneeweis, USN who has collected Mercury through the present space items and gave me permission to use his photos. Also to Robert and Joan Renfree for support over the years. To Daniel Utecht who gave me advice on how to print this book. Most of all, the author wants to thank his wife Beth, who patiently put up with him during the writing of the all editions of this book, and is the inspiration for the element Beccium used in the student's homework. The author would like to also thank Ben Bova and Rick Wilber for their advice and friendship over the years.

About the author

Jeff Mitchell graduated from the Massachusetts Institute of Technology where he studied Electrical Engineering and Physics. His thesis was on the X-ray binary star system AM-Herculus, feel free to read it at the MIT library. He has taught Guidance, Navigation and Control of Spacecraft, Orbital Mechanics, Space Power Systems, Space Environment, and Rocket Propulsion at the Florida Institute of Technology, FIT. This makes him a real Rocket Scientist from MIT. He worked for over 28 years at the Kennedy Space Center helping launch the Space Shuttle and its important scientific payloads, along with thousands of other dedicated workers. His last job was as an astronaut representative, which means he received a lot of the same training as an astronaut. (Ask him about the $250,000 tool he used). As his wife said, we have to get off the planet, this is more important than getting rich. This book will not get him rich.

Chapter 1, Energy Storage

Before we start talking about energy storage, a brief review of various relevant topics is essential, such as units and electrical engineering simple concepts. After all this is about power systems.

Units

Everything in the physical sciences can be expressed in one of the following five quantities: mass, length, time, charge, and temperature. This book uses the System International (SI) system of units for these quantities, mostly Meter Kilogram Seconds (MKS) or Centimeter Gram Second (cgs). Appendix A gives a derivation from various SI units to others (i.e. a Pascal is a Newton per square meter). Appendix B has common constants that will be used throughout this book.

Electrical Engineering Review

Since an Electrical Engineering (EE) degree is not required to learn this material, a quick review of EE things is helpful for those of you who majored in something else.

V is Voltage in Volts. This is a potential, a push.

I is Current in Amperes or Amps. This is the number of positively charged particles passing a point in a second. If the particles are negatively charged, the current flows in the opposite direction of the negatively charged particles movement. Current flows from the positive terminal to the negative terminal, electrons flow from the negative terminal to the positive terminal.

R is Resistance in Ohms. This is the resistance to current flow.

P is Power in Watts, it is voltage times current.

Hz is Hertz otherwise known as cycles per second, an alternating current term.

Ohms law relates these three as follows: V=I times R or

$$V=IR \qquad\qquad 1\text{-}1$$

Power is the voltage times the current or

$$P=VI \qquad\qquad 1\text{-}2$$

Or, using Ohms law

$$P=IxIxR \text{ (I squared R) or } P=(VxV)/R \text{ (V squared over R)} \qquad 1\text{-}3$$

The load is the device that does work by consuming the power. At maximum power the resistance of the load is equal to the internal resistance of the device generating the power.

There are different types of voltages when you deal with real devices, such as open circuit voltage (Voc) versus load voltage (V_L). Voc means that there is no load on the power supply, whereas V_L is the voltage with a load connected. V_L is always less than Voc.

Voltage and resistance add when connected in series. For voltage, just think of your two or three cell flashlight, you push those batteries in one after the other, where the tip of one touches the end of the other. Two 1.5 volt batteries in series like that gives you a 3 volt flashlight.

When two resistors are connected in parallel the total resistance is not a mere addition, it is the following equation.

$$R_t = (R_1 \times R_2)/(R_1 + R_2) \qquad\qquad 1\text{-}4$$

Currents add in parallel.

This is the symbol for a resistor:

This is the symbol for a battery:

This is the symbol for a diode. Current only flows in one direction, from plus to minus.

So far we have only been talking about Direct Current (DC). Alternating Current (AC) is used commonly on Earth and to a lesser extent in space. A wire moving through a magnetic field will produce a current in the wire. If you make a loop of wire and rotate it, you get periods where the wire cuts through the magnetic field, producing voltage and a resulting current, and periods where the wire is parallel to the field and no current is produced. This makes AC. AC is a sinusoidal change in voltage vs. time. One wire in the field produces single phase AC, figure 1.1a.

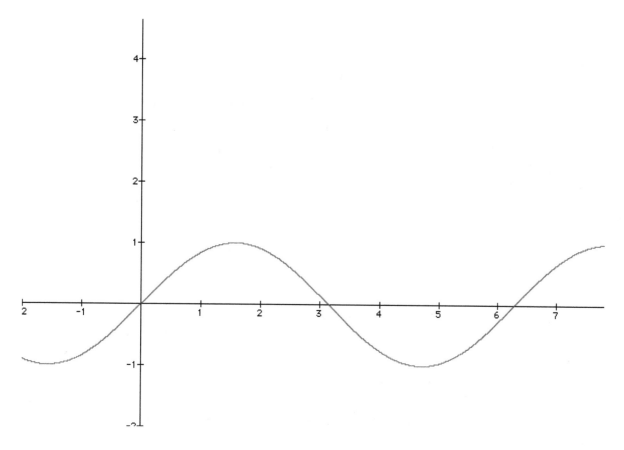

The Y axis is Voltage, the X axis is time.

Figure 1.1a single phase AC

When we put three wires, each 120 degrees apart and rotate them in that same field we get 3 phase AC, figure 1.1b.

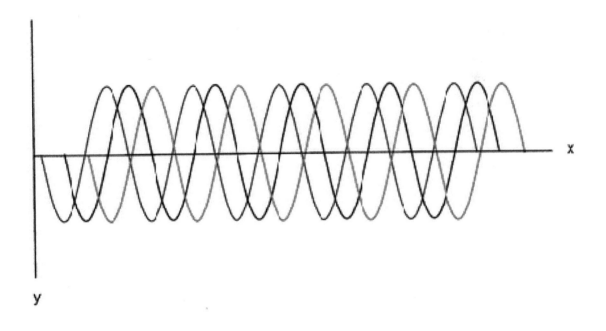

Figure 1.1b three phase AC

AC was used on the Space Shuttle for many things, such as the Payload Retention Latches, the motors on the Payload Bay doors, the External Tank doors, and the vacuum cleaner at 400 Hertz (Hz), 115 Volts 3 phase. Hertz is a measurement of cycles per second. In figure 1.1a above, one cycle is from when the voltage goes from zero to positive to negative and back to zero. In the figure it goes from the x axis equals zero to approximately 6.3. AC was used on the Spacelab Program for their vacuum cleaner, avionics fans, water pumps, and for the CRTs (Cathode Ray Tubes, i.e. computer monitors) of the Data Display Units. AC in your home is at 115V single phase 60 Hertz. On Spacelab it was 440 V, 3 phase, 50 Hz, since Europe uses 50 Hz and Spacelab was built in Europe. AC was used on the SPACEHAB program as well (3 phase), but it was built in the United States so it used 60 Hz. The three phases are 120 degrees apart, that is, phase A starts at time 0, phase B starts when phase A is at 120 degrees along it's sine wave (in figure one it would be at 1), and phase C starts when phase A is at 240 degrees (figure 1.1b). Why use three phase vs. single phase? Well, pumps, fans, and other rotating devices rotate more easily with 3 phase AC. Also, if a fan or pump is rotating and loses one phase, it will continue to rotate. With single phase, once you lose the phase the rotating machinery will stop. There are two different ways to connect 3 phases, Delta and Y phasing (figure 1.2).

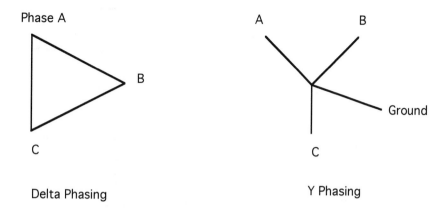

Figure 1.2 Delta and Y phasing

On single phase, in your house for instance, you get power by putting the load across the phase and ground. In three phase you get power by putting the load across two or three phases. On Y phase you can also get power by putting the load across any phase and ground. The United States Navy uses Delta phasing because their personnel are on a metal ship and if one of them touches one phase accidentally, and is grounded to the metal ship, they would get electrocuted. The Spacelab and Shuttle Programs used Y phasing, which makes no sense because astronauts tend to hold onto metal handrails, so if their other hand hits one of the phases, the electricity would have a good path to ground through the heart of the astronaut. This is not good. The International Space Station (ISS) fortunately uses DC.

AC is easily converted to different voltages by using a transformer. A simple transformer is a iron square ring with two coils of wire wrapped around the opposite sides. One side is the input, and the other the output (figure 1.3). The input has many windings about the core. Let's say seven. The output side has more windings, let's say fifty-six. So, if we put one volt AC on the input side, the alternating current produces an electromagnetic field which alternately builds and collapses. This field produces another field on the other end of the core which induces a voltage. The net result is that the voltage is amplified by 56 divide by 7 equals 8 times. We get 8 volts out the other side. One volt in gives 8 volts out the other side.

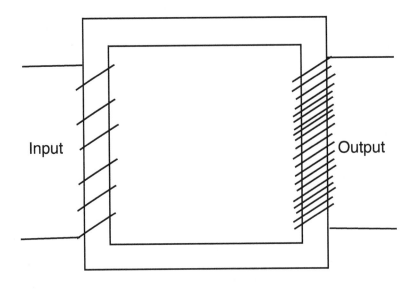

Figure 1.3 Transformer

Let's get back to DC. DC has to be converted to AC in order to transform it to a different voltage. How do we do that? Well, think about old car radios--before we had transistors we had vacuum tubes, which used AC to heat their filaments. So, in the car the batteries 12 volt DC voltage was interrupted by using a buzzer, and then that square wave voltage (figure 1.4) was sent into a transformer to be turned into the proper voltage. A square wave looks like alternating current to a transformer. (In fact, a square wave can be mathematically produced by the addition of a large number of sine waves using a Taylor series polynomial expansion, or something like that). Today we use complex solid state devices to make a square wave to send to a transformer. These devices that convert DC to AC are called Inverters. The most common ones that people use plug into their car 12 VDC outlet and output 115 VAC to recharge their appliances (laptop computers, phones, etc.) If you then take that AC and turn it into DC at that new DC voltage you have a device called a Converter.

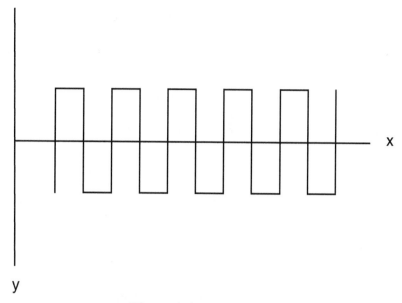

Figure 1.4 square wave

Energy Storage

One component of most Space Power Systems is the energy storage device. Most satellites use batteries to ensure that they can continue transmitting during the dark period (eclipse) of their orbits. Satellites also store mechanical energy with Momentum wheels. Future demands for energy storage will include thermal storage.

Batteries

History

The earliest batteries, believe it or not, are thousands of years old. In 1936, near Baghdad, explorers unearthed a tomb belonging to the Parthian Empire, at about 200 BCE (Before Common Era or Current Era). They found a vase or jar, sealed with pitch at the opening on the top. An iron rod came out the top, surrounded by a copper tube. Replicas made of this jar, once filled with acid, produced about 1.2 to 2 Volts. They speculate that many of these batteries were used to electroplate gold onto silver articles.

In 1800, Alessandro Volta made the first "pile" battery. He piled up up layers of silver, paper or cloth soaked in salt, and zinc. Many triple layers were made into a tall pile, without paper or cloth between zinc and silver. The layers were stacked on each other until the desired voltage was reached. Fast forward to today. The French word for battery is 'pile' (English pronunciation "peel".)

In 1820 John Frederich Daniell, constructed a cell that put out more current for a longer time. He had a jar, with a copper plate on the bottom. On top of that was a copper sulphate solution that extended halfway up the remaining distance toward the top plate of zinc. On top of the copper sulphate was a zinc sulphate solution, which was less dense and floated above the copper sulphate. The zinc electrode was on the top. The amazing thing was that it powered telegraphs, telephones, and rang doorbells for over 100 years.

All batteries operate in the same fashion. A chemical reaction produces electricity: i.e. chemical energy is converted to electrical energy. Batteries come in two flavors, recharge-able and non-rechargeable. The former is called a secondary battery, and the latter is called a primary battery. Both have the same basic components, two electrodes in an electrolyte. The secondary battery may also have a separator which prevents the deterioration and eventual shorting of the battery.

Figure 1.5 shows a typical battery called a Daniell cell. The negative electrode of a battery is called the anode and the positive electrode is called the cathode. The anode is where oxidation takes place, and the cathode is where reduction takes place. Oxidation is the losing of electrons, reduction is the gaining of electrons. An anion is a negatively charged ion and a cation is a positively charge ion. This anode is made of Zinc (chemical symbol Zn) and the cathode is made of Copper (Cu). When the two electrodes are hooked together with a conducting wire, electrons leave the atoms in the anode, oxidizing the anode. The Zn loses two electrons which generates cations that go into the electrolyte solution (Zn^{++}). The electrons flow to the copper cathode where they combine with the Cu^{++} cations in the solution, reducing them to form copper, which plates out on the cathode. The reaction continues until either the Zinc anode disappears or the Cu^{++} ions are depleted. The reaction can also stop if the copper plates out and pierces the membrane or, in batteries without a membrane, connects with the anode, causing a short.

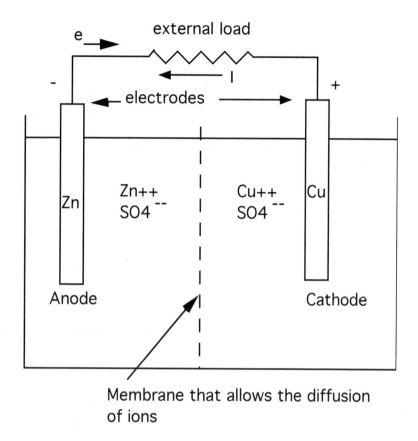

Figure 1.5 Daniell cell

How do we know whether two chemicals can make a battery? Chemists like to attribute part of a cell's standard potential, $V°$ to each of the half-reactions that go to make up the cell reaction. Chemists define the half reactions of Hydrogen as follows:

$$H_2(g) \rightarrow 2H+ + 2e- \quad V° = 0$$

Looking at the table 1.1 we see some typical half reactions and their potentials. This will enable us to see if a cell will work, and what its potential would be.

Table 1.1 half reactions.

Half Reaction (s = solid, g = gas)	Potential (Volts) V^0
$Ag(s) \rightarrow Ag^+ + e^-$	-0.799
$Al(s) \rightarrow Al^{+3} + 3e^-$	+1.66
$2Br^- \rightarrow Br_2 + 2e^-$	-1.06
$2Cl^- \rightarrow Cl_2(g) + 2e^-$	-1.36
$Cu(s) \rightarrow Cu^{++} + 2e^-$	-0.337
$2F^- \rightarrow F_2(g) + 2e^-$	-2.87
$Fe(s) \rightarrow Fe^{++} + 2e^-$	+0.44
$H_2(g) \rightarrow 2H^+ + 2e^-$	0
$H_2O(l) \rightarrow 1/2 O_2(g) + 2H^+ + 2e^-$	-1.23
$4H_2O + Mn^{++} \rightarrow MnO_4^- + 8H^+ + 5e^-$	-1.51
$Hg(l) \rightarrow Hg^{++} + 2e^-$	-0.85
$2I^- \rightarrow I_2 + 2e^-$	-0.536
$Li(s) \rightarrow Li^+ + e^-$	+3.05
$Mg(s) \rightarrow Mg^{++} + 2e^-$	+2.37
$Na(s) \rightarrow Na^+ + e^-$	+2.71
$Sn^{++} \rightarrow Sn^{+4} + 2e^-$	-0.15
$Zn(s) \rightarrow Zn^{++} + 2e^-$	+0.763

For the Daniell cell, let us look at the half reactions.

On the left side, we have Zinc losing two electrons. On the right side, Copper gains two electrons. From table 1.1 the half reactions are:

$$Zn(s) \rightarrow Zn^{++} + 2e^- \qquad\qquad +0.763 \text{ V} \qquad\qquad 1\text{-}5$$

$$Cu(s) \rightarrow Cu^{++} + 2e^- \qquad\qquad -0.337 \text{ V} \qquad\qquad 1\text{-}6$$

The entire reaction is

$$Zn(s) + Cu^{++} \rightarrow Zn^{++} + Cu(s) \qquad\qquad 1\text{-}7$$

All we have to do is add both equations:

$$Zn(s) \rightarrow Zn^{++} + 2e^- \ +$$

$$\underline{Cu(s) \rightarrow Cu^{++} + 2e^-} \ =$$

$$Zn(s) + Cu(s) \rightarrow Zn^{++} + 2e \ Cu^{++} + 2e^- \qquad\qquad 1\text{-}8$$

But that isn't equation 1-7. So, we have to reverse equation 1-6, which yields:

$$Cu^{++} + 2e \rightarrow Cu(s) \qquad\qquad +0.337 \text{ V} \qquad\qquad 1\text{-}9$$

Notice that V^0 is now a positive number. Now we add.

$$Zn(s) \rightarrow Zn^{++} + 2e^- \ + \qquad\qquad +0.763 \text{ V}$$

$$\underline{Cu^{++} + 2e \rightarrow Cu(s)} \ = \qquad\qquad +0.337 \text{ V}$$

$$Zn(s) + Cu^{++} \rightarrow Zn^{++} + Cu(s) \qquad\qquad +1.100 \text{ V}$$

The electrons are on both sides, so they cancel. Since the final voltage is positive, this reaction can be used to make a battery. If was negative, you couldn't use it as a battery.

Battery Selection Process for Space Missions

There are four considerations in the battery selection process: Physical constraints, Electrical Performance, Environmental Conditions, and Special Considerations.

Physical Constraints

Weight--Obviously lead acid batteries are much heavier than flashlight batteries.
Size--A D cell and a AAA cell are both rated at 1.5 volts, but are of different size.
Shock and Vibration--Some batteries are more rugged than others.
Configuration--How many plates, how many cells, location of terminals, etc.

Electrical Performance

Voltage Limits--These are the upper and lower voltages at which the battery operates.
Current Capacity--How much current can it produce.
Stability--Is it explosive? Will it deteriorate when stored for long periods?
Discharge Rate--How fast does it discharge under a load or in storage?
Load Profile--Voltage versus time under a load.
Recharge Requirements--Does the battery need a large current for a short time or a medium current for a medium time?

Environmental

Storage times--For both dry and wet life. Wet life starts from when the battery is activated by adding the electrolyte.

Operating Temperature, Humidity, and Pressure--These are the conditions at which the battery can be operated. The term Operating Temperature is also used sometimes to indicate the temperature of the battery when under a load producing power.

Special Considerations

Is it a Primary or Secondary battery?
Is the battery automatically activated during the course of the Space Mission, or is it manually activated at the launch pad?
How much does the battery cost?
Is the battery explosive or corrosive? How safe is it?

Selections for Space Missions

The factors for using a battery for Space Missions are pretty obvious. Since the cost of lifting a payload into orbit is very high, batteries should be very light. Also, the volume should be small so the total vehicle size and hence weight will be small. Therefore high power per mass and power per volume batteries are highly desirable. In this book, power per volume (W/m^3) is called power density. Some books will call this volume power density or volume specific power. This book will not use those terms. In this book, power per mass is called specific power. This is not to be confused, but often is, by energy density, specific energy, and other terms. Remember, energy is power times time. Energy, a Joule (J), is a Watt times a second (W-s). So energy density is a Ws/m^3, not the same as power density. By the way, the electrical meter on your house reads in kilowatt-hours (kW-hr). You don't get a power bill from your electric company, you get an energy bill. Be sure to tell them that when you pay your next bill.

Description	Units	Nomenclature
Watts per volume	W/m^3	power density (or volume specific power)
Watts per mass	W/kg	specific power
Energy per volume	J/m^3 ($W\text{-}s/m^3$) also Whr/liter	energy density
Energy per mass	J/kg ($W\text{-}s/kg$) also Whr/kg	specific energy

Although Lithium batteries have the highest ratios, they are potentially explosive, so were considered undesirable for space batteries for many years. However, recently advances in Lithium battery construction have led the them being used in satellites. The European GPS system, Galileo, uses Lithium-ion (LI-ion) batteries. The discharge characteristics of various batteries are interesting. The silver zinc battery has a nearly constant voltage over 90% of its discharge characteristic. This is highly desirable and one of the primary reasons it was chosen for the Inertial Upper Stage (IUS). The higher voltage at the beginning of the discharge is called the peroxide voltage and cannot be eliminated. On the first flight of the IUS, the ground crew found that the IUS computers were very sensitive to voltage dips which occurred when the load was applied. For subsequent missions the batteries were increased from 100 to 140 amp hours and given a 50% preflight discharge. This put the voltage in the stable range so the computers would operate properly.

Nickel-Cadmium (Ni-Cd or Ni-Cad) batteries are used on some spacecraft. These batteries have a property known as memory. If a Ni-Cd battery is overcharged a chemical change occurs that makes the discharge of the battery more difficult. If the battery is discharged to only 80% of its capacity and then is recharged to 100%, the battery is chemically reset so that 80% is now 0%. The first 20% of discharge (from 100% to 80%) will discharge easily, the remaining 80% will remain in a difficult discharge state. These batteries must be reconditioned periodically. Reconditioning is usually done just prior to solar eclipse to ensure maximum reliability and charge. In Geosynchronous Earth Orbit (GEO) satellites, Ni-Cd batteries are usually operated at 50% Depth of Discharge (DOD). Complete discharge will cause expansion and contraction of the electrodes, gradually weakening them and reducing lifetime.

The original batteries for the Hubble Space Telescope (HST) were Ni-Cd. But the HST was colder than expected, which lead to more power being used for the heaters. The memory of Ni-Cd batteries requires them to be discharged completely more often than Nickel-Hydrogen (Ni-H) batteries. The six Ni-Cd batteries originally designed for HST had a capacity of 69 amp hours and had to go through 16 charge-discharge cycles a day. Ni-Cd batteries lose 40% of their capacity under these conditions after two or three years. This means that periodic replacement is necessary, which is exacerbated by the heating problem. One or two more Shuttle missions would be needed just to change the batteries. After the Challenger accident there was a crash study made of Ni-H batteries as a replacement for the Ni-Cd for the HST. Ni-H are usually used in GEO satellites, and therefore have never had to survive the rapid charge-discharge cycles of Low Earth Orbit (LEO). Since Ni-H batteries have higher energy/mass than Ni-Cd batteries more can be put on the HST for the same mass as Ni-Cds and more energy is produced. The Ni-H batteries can handle twice the peak power demand of the Ni-Cd batteries, allowing more instruments to run, or more frequent shifts in attitude. The HST uses electricity to for attitude change, there are no thrusters because thruster exhaust would deposit on the lenses, making them blurry and useless.

Ni-H batteries (figure 1.6 and 1.7) use Hydrogen gas at 82.7 bar (1200 psi). Therefore this is a cylindrical battery with hemispherical ends.

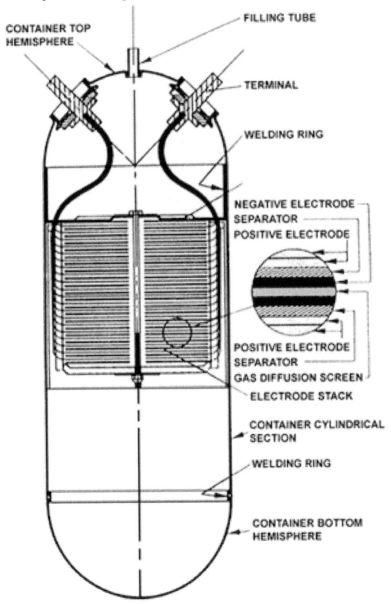

Figure 1.6 Nickel-Hydrogen battery (Courtesy NASA)

Figure 1.7 Hubble Space Telescope Ni-H batteries (Courtesy NASA)

Newer batteries are being invented all the time, many of these will be adapted for Space Missions. Nickel-metal Hydride (NiMH) batteries are similar to Ni-H except the hydrogen is contained in a hydride, thus it is a low pressure device. It has almost twice the energy density of Ni-Cd and are equivalent to Ni-H. Plastic conductors have been invented and proposed for the electrodes in batteries. With a lighter weight and the ability to be formed into odd shapes, plastic batteries could be developed to fit into leftover spaces in spacecraft. Table 1.2 compares the various properties of batteries. The first four batteries are primary batteries, the rest are secondary.   Note that there is no correlation between energy per mass (specific energy) and energy per volume (energy density). Decisions have to be made to pick the optimum of both for a mission.

Table 1.2 various battery characteristics.

Battery	Anode	Cathode	Energy per mass, Specific Energy Whr/kg	Energy Density Whr/liter	Voltage
Mercury (primary)	Zn	HgO	100	460	1.35
Silver Oxide (primary)	Ag	Ag_2O	130	500	1.6
Zinc Air (primary)	Zn	O_2	470	1480	1.65
Lithium Magnesium (primary)	Li	MnO_2	230	550	3
Lead Acid	Pb	PbO_2	35	60	2.1
Ni-Cad	Cd	NiO_2	40	80	1.24
Nickel Metal Hydride	MH	NiO_2	60	140	1.2
Nickel Hydrogen	H_2	NiO_2	75	60	1.55
Silver Zinc	Zn	AgO_2	130	500	1.86
Silver Cadmium	Cd	AgO	55	100	1.1
Lithium Ion	C	$LiCoO_2$	100	250	3.6
Lithium Organic	Li	Mn_2O_4	120	230	3
Sodium Sulfur	Na	S	160	110	2

Sizing

Now let us work a problem and figure out the size of a battery used in a GEO space-craft. First we will define some terms. The total capacity of a battery is given in Ampere-hours or Amp-hours. A 100 A-h battery can supply 100 amps for 1 hour. The Depth of Dis-charge (DOD) is the percent of battery capacity used in discharge, i.e. supplying power. An 80% DOD means there is 20% capacity remaining. Watt-hours are the stored energy of the battery, which is A-h capacity times discharge voltage. The charge rate is the rate at which the battery can accept charge. This can vary depending on what kind of charge you are giv-ing the battery (trickle charge is a lower amperage than a recharge). The average discharge voltage is the number of cells in series times the average voltage of one cell. Finally, most space batteries operate at about 1.25 volts per cell.

DOD affects the life and mass of a battery system. Repeated charge and recharge cy-cles, as occurs in a LEO spacecraft reduce battery life. A spacecraft with a period of 100 minutes or less will be discharged and charged roughly 16 times a day, or 6,000 times a year. Spacecraft in GEO have two eclipse seasons a year, with an eclipse varying from seconds to 72 minutes. The less DOD used means that battery mass is being wasted by not using the capacity. Too much DOD reduces battery life. You can size a battery using the following equation.

DOD = (Energy used during eclipse)/(Total battery stored energy) or

DOD = (Load power times discharge time)/(Capacity times battery average discharge volt-age)

The charge rate drives the battery size. One arbitrary rule is:

Charge rate = Capacity/15 hours

Note: Look at the instructions for a Ni-Cd battery device that you have in your home. The instructions may say charge for 16 hours before use. So its charge rate = Capacity/16 hours. Both of these numbers are conservative and really depend on the battery. Some bat-teries can recharge much faster.

Example 1-1: Size a battery for a 1500 W payload in GEO. Bus voltage is 28 V. Load duration is 1.2 hours max (72 minutes). Energy density is 33 W-hr/kg. Cell voltage is 1.25 V. Maximum DOD is 70%.

Number of cells needed = 28V/1.25 = 22.4 You can choose 22 cells and get 27.5 volts or 23 cells for 28.75 V. Let's use 27.5 V.

$$\text{Total capacity} = \frac{1500 \text{ W x } 1.2\text{h}}{0.70 \text{ DOD x } 27.5 \text{ V}} = 93.5 A - h$$

Battery Stored Energy = 93.5 A-h x 27.5 V = 2571 W-h

Battery mass = 2571W-h/33 W-h/kg = 77.9 kg

Batacitor/Supercapacitor/Ultracapacitor (we'll use ultracapacitor)

These are all the same devices, the industry hasn't standardized the name yet. Basically it is a combination of a battery and capacitor. Both use an electrolyte. In a battery, ions are moved into and out of the electrodes by chemical reactions. In a ultracapacitor an electric field moves the ions to and from the surfaces of the electrode. Because the ions are on the surface, you can charge and discharge one quickly, in seconds as opposed to hours for a battery. Currently the ultracapacitor can store about 5% as much energy as a Li-ion battery. There are companies that are claiming they can make a device that will store twice as much energy and deliver 10 times as much power as a regular battery. Once these are developed, these will be ideal for space applications.

Momentum Wheels--Mechanical Storage

The attitude control systems on some spacecraft consist of momentum wheels (Momentum Wheel Assemblies MWAs). MWAs are spinning wheels storing momentum. Sun and Earth sensors feed data to the onboard attitude computers which send signals to magnetic torquers. These activate and the magnets slow the MWAs and give a resulting torque to the spacecraft. On the Satcom satellites, for instance, this method controls only pitch, propellant is used for yaw and roll. On the Tracking and Data Relay Satellite (TDRS), two reaction wheels are used. Pitch and roll errors are measured by Earth sensors which control reaction wheel speeds to control roll and pitch pointing, respectively. Yaw control is provided by the momentum bias of the reaction wheels with periodic ground generated commands to the thrusters for removal of accumulated momentum. You can only speed up a wheel to a maximum rate, then you have to slow it down and use thrusters to correct the attitude. The exact energy storage capabilities vary with each spacecraft. Note that if the momentum wheel is completely stopped, you've achieved a 100% DOD. MWAs can be spun up and down with no reduction in storage capacity.

Figure 1.8 Small momentum/reaction wheel developed for NASA's Small Explorer (SMEX) program. (Courtesy NASA)

Future Storage

Space Platforms, such as the Space Station, have their own energy storage requirements since they will also have periods of eclipse. The two basic ways to compare the energy storage systems are the efficiency which is simply energy out divided by energy in, and the storage system mass per unit energy stored. The Depth of Discharge, DOD, which is the fraction of total energy stored that can be routinely taken from storage without seriously affecting the life of the storage system, varies depending on the system.

For advanced power systems using solar Brayton cycles (a Brayton cycle is one where heat expands a gas, which turns a turbine to generate electricity), solar energy could be stored by a fluid or salt using the latent heat of fusion. The stored heat energy would be used when the Space Platform was eclipsed to keep the generators running. Efficiencies of up to 90% are possible. This is what was being planned for the original International Space Station (ISS), and we will discuss this more in a ISS chapter.

Solar arrays can recharge Ni-Cd or Ni-H batteries. These efficiencies are about 65%. A regenerative fuel cell system is possible, where the solar panels would electrolyze the water produced by the fuel cell, to make hydrogen and oxygen. (The fuel cells would be used for peak power loads). This system could have an efficiency of approximately 60%. Finally, the arrays could be used to spin momentum wheels for storage. During the eclipse the motors that spun the momentum wheels would be used as a generator. Near term efficiency of 80% and far term of 90% are likely. NASA was developing a flywheel energy storage (FES) system for the International Space Station. See the chapter on ISS. The Station's solar arrays currently provide power for the station, part of that electricity is used to charge the batteries. When the Station enters the shadow of the Earth, the batteries supply the power to the station through the Battery Charge/Discharge Unit (BCDU). The FES would have replaced the batteries and the BCDU. The flywheels would store more energy than the batteries, thus allowing extra electrical capacity for the experiments. Batteries have a limited amount of charge/discharge cycles. Flywheels can be charge/discharged virtually forever. Therefore they last longer and should save 100's of millions of dollars over the life of the station. Testing at JSC was to start in 2001 with a planned launch to the Space Station in 2004. However, this was cancelled by the Bush administration to save cost in 2001. Note that this is not a sound business decision considering the 100's of millions of dollars that would have been saved.

Problems

1. A battery consists of 6 cells hooked in series as shown.

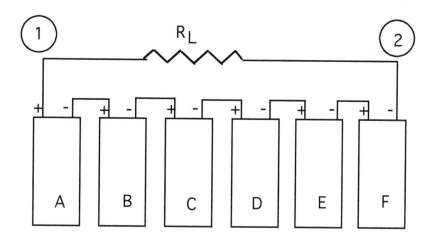

Each cell is hooked to a voltmeter and the following voltages are recorded.

VA = 1.25V   VB = 1.23V   VC = 1.22V   VD = 1.21V   VE = 1.21V   VF = 1.28V

a) What is the total voltage measured across 1 and 2?

b) Which way do the electrons flow, from 2 to 1 or 1 to 2?

c) If the Load Resistance is 3 ohms, how much current flows?

d) How much power is produced by the load?

2. I have rechargeable batteries A that mass 12 grams per watt output. I have rechargeable batteries B that put out 0.5 amps at 1.12 volts and mass 6.2 grams. Assuming size is unimportant, which do I choose for a space mission and why?

3. Determine the resistance between points A and B.

a)

5 ohms   10 ohms

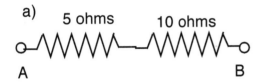

A   B

b)

100 ohms

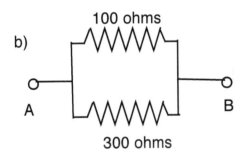

A   B

300 ohms

c)

10 ohms   100 ohms

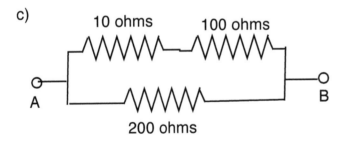

A   B

200 ohms

4. What is the voltage, current, and power in the following circuit, assuming it drives a load.

1.4 V 1.0 A   1.4 V 1.0 A

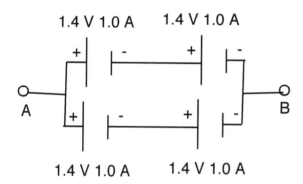

1.4 V 1.0 A   1.4 V 1.0 A

5. Using the example 1-1 on sizing a battery, only now use Ni-H properties:

Size a Ni-H battery pack for a 1800 W payload in GEO. Bus voltage is 28 V. Load duration is 1.2 hours max (72 minutes). Maximum DOD is 70%.

6. How many Ni-H batteries do I hook up in series to make a string, and then how many strings do I need in parallel to get >= 28 Volts and >= 2.4 Amps, assuming that the batteries are rated at 600 mAh (milli-Amp-hours) and they will be used for 1 hour.

7. What will the V_0 be for each case below and which of the following reactions might be used as a source of electrical energy?

a) $I_2(s) + 2Br^-(aq) \rightarrow 2I^-(aq) + Br_2(l)$

b) $2Ag(s) + Zn^{++}(aq) \rightarrow 2Ag^+(aq) + Zn(s)$

c) $Cl_2(g) + 2Br^-(aq) \rightarrow 2Cl^-(aq) + Br_2(l)$

Chapter 2, Fuel Cells

History

In 1839 William Robert Grove invented the "gas battery" as he called the fuel cell. This was just 39 years after Alessandro Volta invented the battery. Grove's materials were unstable and the invention didn't take off. In 1889 Charles Langer and Ludwig Mond, coined the term "fuel cell" as they attempted to make the first practical fuel cell using air and coal gas. More attempts were made in the early 1900s to make coal or carbon fuel cells, the pesky internal combustion engine temporarily was more efficient and stopped development.

Francis Bacon developed the first successful fuel cell device in 1932, with a hydrogen-oxygen cell using alkaline electrolytes and nickel electrodes - inexpensive alternatives to the catalysts used by Mond and Langer. There were technical problems and in 1959 Bacon's company had a practical 5 kW fuel cell. In the late 1950's NASA funded research for fuel cells.

Fuel Cells

Fuel cells have been used on the Gemini, Apollo, and Space Shuttle vehicles. For space missions they produced power and generate water as a by-product. They are very efficient, and very expensive.

Fuel cells can be classified by type of fuel, temperature, and electrolyte. Since basic cell configuration is controlled by the type of fuel, this is the most common way. When classified by temperature, they can be split into high, medium and low temperature cells. High temp cells run from 500 to 650 degrees centigrade with molten salt electrolytes. Medium temp cells operate up to the boiling point of the electrolyte. Low temperature cells are mainly what we are concerned with in space power systems. The electrolytes are either basic or acidic. These can be subdivided into solid, liquid or contained electrolytes. The most common contained electrolytes are proton exchange membrane.

Finally, classification by fuel leads us to three states; solid, liquid, and gas. Coal or metals are solids, alcohol a liquid, and hydrogen is a common gas used in fuel cells.

Fuel cells offer some distinct advantages over other methods of conversion. Any power plant that burns fuel and produces power is limited by the Carnot efficiency,

$$\eta_C = 1 - \frac{T_C}{T_H}$$

In practice, other inefficiencies creep in and the total efficiency is, at most, 40%. Fuel cells do not burn fuel, and are thus not limited by the Carnot theoretical maximum. They can achieve 70-80% efficiency.

Figure 2.1 Apollo Command Service Module Fuel Cells (Courtesy Lieutenant Commander Scott B. Schneeweis, USN http://www.spaceaholic.com)

How fuel cells work

Fuel cells are essentially a primary battery where the fuel and oxidizer are stored external to the battery. It is still a combustion reaction, the fuel molecules lose electrons (oxidation) and the oxidizer molecules gain them (reduction). In a heat engine, this is done by mixing and burning the fuel and oxidizer. In a fuel cell the fuel and oxidizer are kept from mixing and the electrons are transferred through the electrical load. Figure 2.2 shows the basic schematic of a fuel cell.   *** need to label anode and cathode in this diagram

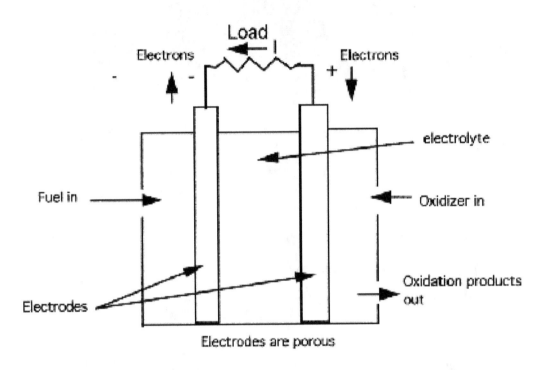

Figure 2.2 Fuel cell

For any steady flow or closed system in temperature and pressure equilibrium with its surroundings we recall from thermodynamics that the maximum useful work is given by:

$$-\bar{W}_{max} = \Delta \bar{G} \ [kcal] \hspace{4cm} 2\text{-}1$$

$\bar{G}$ is the Gibbs function, $\bar{H}\text{-}\,T\bar{S}$

Where $\bar{H}$ is enthalpy and $\bar{S}$ is entropy. (A bar over the letter indicates a property in simple units, such as kilocalories. The chemists use the convention of no bar indicates a property in per moles. Thus H bar would be in kilocalories and H would be in kilocalories per mole).

If we assume that the only useful work we can do is electrical work, then we can write this in a per mole fashion as:

$$-W_{max(el)} = \Delta G \ [kcal \,/ \, mole] \hspace{3cm} 2\text{-}2$$

The electrical work the cell produces is the quantity of charge that leaves the cell (current) multiplied by what is pushing it out of the cell, the potential difference of the cell, V. The quantity of charge is the number of moles of electrons in the reaction times the number of coulombs per mole of electrons. Therefore:

$$W(el) = -nFV \; [kcal/mole] \qquad\qquad\qquad 2\text{-}3$$

where n is the number of moles of electrons and F is a Faraday, a charge equal to Avogadro's number of electrons, 23.06 kcal/volt-mole of electrons. Combining these two equations we get

$$-W_{max}(el) = \Delta G = -nFV \; [kcal/mole] \qquad\qquad 2\text{-}4$$

V is the reversible potential of the cell, which is the open circuit voltage. We use a superscript degree sign to designate thermodynamic quantities in there standard states, usually at 1 atmosphere. V° is the standard cell potential (also called the standard emf [electromotive force]) corresponding to the standard free energy change, ΔG°. Thus:

$$\Delta G^0 = -nFV^0 \; [kcal/mole] \qquad\qquad\qquad 2\text{-}5$$

If we go through a detailed analysis of the chemistry of the fuel cell we'll come up with the following equation:

$$\Delta G^0 = \Delta G - RT \ln Q' \qquad\qquad\qquad 2\text{-}6$$

where R is the universal gas constant which is equal to 1.987 cal/deg/gm-mole of reaction, T is the absolute temperature in Kelvin, and Q' is the actual concentration (or activity) ratio of the products to the reactants. (Activity is a measure of the "effective concentration" of a chemical. The activity of a pure substance such as a solid or liquid is 1). We can get rid of ΔG° to obtain:

$$-nFV^0 = -nFV - RT \ln Q' \qquad\qquad \text{or,}$$

$$V^0 = V + (RT/nF)(\ln Q') \qquad\qquad\qquad 2\text{-}7$$

This is the <u>Nernst equation</u> and it will enable us to determine the open circuit voltage of a fuel cell. Recall from chapter one the discussion on half reactions and table 1.1. Let us look at hydrogen and oxygen producing water. This reaction was used on the Space Shuttle fuel cells.

$$H_2(g) + 1/2 O_2(g) \rightarrow H_2O$$

Looking at the half reactions in table 1.1.

$$H_2(g) \rightarrow 2H^+ + 2e^- \qquad\qquad V° = 0 \qquad\qquad 2\text{-}8$$
$$H_2O(l) \rightarrow 1/2 O_2(g) + 2H^+ + 2e^- \qquad V° = -1.23$$

But, if we add those two together, we don't get the right equation. We have to reverse the second equation, and this reverses the sign of $V°$.

$H_2(g) \rightarrow 2H^+ + 2e^-$ $\qquad\qquad$ $V° = 0$
$1/2O_2(g) + 2H^+ + 2e^- \rightarrow H_2O(l)$ $\qquad$ $V° = +1.23$

When you add those two equations, the $2H^+ + 2e-$ are on both sides of the equation, so they cancel, and we get the correct equation

$H_2(g) + 1/2O_2(g) \rightarrow H_2O$ $\qquad\qquad$ $V° = +1.23$

So, as with the battery example in chapter one, for any fuel cell that has a $V°$ that is positive, the reaction can take place.

Let's use the Nernst equation for a hydrogen oxygen fuel cell. With temperature at 25 C, hydrogen supplied at 1 atmosphere, a 30% KOH mixture, air as the oxidizer, and the half reaction from table 2.1 of $V° = 1.23$ V then:

$$V = V^0 - (RT/nF)(\ln Q') \qquad\qquad\qquad 2\text{-}9$$

(The minus sign is because we are looking for V not V^0)

At a 30% KOH concentration the activity (the effective concentration in chemical terms) of water is slightly less than one, but we will use one, this will make the answer conservative. The hydrogen is supplied at 1 atmosphere and therefore has an activity of 1 because the activity is equal to the partial pressure of an ideal gas. The activity of oxygen is therefore 0.21 because that is the partial pressure of oxygen in air (21% = 0.21). Since the Oxygen is O_2 we take the partial pressure to the 1/2 power. Thus the open circuit voltage (also called the reversible voltage) is:

$$V = 1.23 - \frac{(1.987)(298)}{2(23060)} \ln \frac{(1)}{(1)(0.21)^{0.5}}$$

$$= 1.219981 = 1.22V$$

The efficiency of a simple fuel cell

A simple fuel cell is one which is not regenerative, we will talk about regenerative fuel cells at the end of this chapter. The ideal efficiency is the change in free energy (the maximum useful work we can obtain from the cell) divided by the heat of reaction

$$\eta_i = \Delta G / \Delta H = 1 - T\Delta S / \Delta H \qquad \text{2-10}$$

This is the thermal efficiency of the fuel cell alone and does not include the associated subsystems used to, for instance, supply the reactants (the pumps). In terms of the electromotive force of the cell, the efficiency is :

$$\eta_i = -nFV / \Delta H = -ItV / \Delta H \qquad \text{2-11}$$

Where I is the current in Amps and t is the time in seconds that the current flows. In a fuel cell under load the voltage will fall below V to some value Vac (V actual). The reasons are attributed to one or more of the following:

1) The reaction may be hindered at the cathode or anode
2) An undesirable reaction may be taking place at the electrodes or in some other place
3) A concentration gradient may occur in the reactants or the electrolyte (higher concentrations of the electrolyte or reactants)
4) Joule heating due to the Current times Resistance (IR) drop that occurs in the electrolyte

Therefore, these effects give an actual efficiency of:

$$\eta_{ac} = -nFV_{ac} / \Delta H = -ItV_{ac} / \Delta H \qquad \text{2-12}$$

We can define the voltage efficiency as:

$$\eta_v = V_{ac} / V \qquad \text{2-13}$$

And finally, the fraction of the reaction which occurs electrochemically, giving current is called the Faradaic or current efficiency.

$$\eta_F = I / nFN_{fu} \qquad \text{2-14}$$

Where N_{fu} is the total number of moles of fuel reacted per second.

Practical Limits to Performance

Before discussing this, there are two important concepts to be discussed that are general requirements of all fuel cells.

The first, reactivity is closely related to the four reasons discussed earlier for a lower than predicted voltage. To produce the best reactivity the proper stoichiometry must be achieved. That is, we want to have carbon changing to CO_2 and not CO, or else we will not get the most efficient reaction. (2 faradays of electricity per mole vice 4). We therefore need high electrode activity, meaning large current densities.

The reactivity problem has been solved by using porous electrodes to increase the area at the gas-electrode interface, by increasing temperature, by increasing pressure, and by using catalysts.

The second requirement, invariance means that a fuel cell should remain constant throughout its lifetime. This means no corrosion, no side reactions, no change in the electrodes or electrolyte, and no fuel diffusing over to mix with the oxidizer. Catalysts can be made ineffective with extraneous material, and the pores of the gas electrodes can be blown though with gas or drowned with liquid, which causes the electrode to be inefficient. This requirement, invariance, is met by the proper choice of operating conditions and materials used.

As with all real things, solving one problem may cause another. If we raise the temperature to improve reactivity, the electrodes may become involved in the cell reaction, losing invariance.

The losses at the electrodes are referred to as an effect of polarization. This term merely denotes a difference between the theoretical and measured voltage of a cell. There are three flavors, chemical, concentration, and resistance.

Chemical or activation polarization is a surface phenomenon, and the amount of polarization is a factor of how ions are discharged at the electrodes and the rate at which they are discharged. The fuel must bond with the electrode and then this bond must be broken, releasing electrons, and the fuel ion goes into the electrolyte. This all takes energy, and this must be subtracted from what is available. To reduce chemical polarization, the gas diffusion electrode has been used. It maximizes the interface of electrolyte-electrode-gas. As a conductor with pores varying from 100 to 10 microns in diameter the reactants and products easily enter and exit the electrode. This creates large surface areas per unit area. Figure 2.3 shows a schematic of one. This prevents drowning, mentioned earlier.

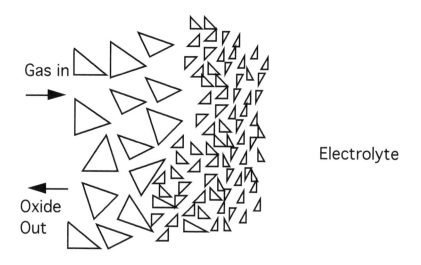

Figure 2.3 Porous electrode

Concentration polarization is caused by the fact that the chemicals can't maintain the initial concentration of the fluid. This creates a back emf (electro-motive force) which is in the opposite direction of the voltage the cell would produce under completely reversible conditions. This polarization can be reduced by increases in temperature, which accelerates ionic diffusion within the cell, or by stirring the electrolyte. (The latter was used in an electric car design. The electrolyte was pumped through all the batteries and performance was improved dramatically). This loss is a very big problem in space because no gravity equals little ion drift.

Resistance polarization occurs because as the chemical reaction goes on in the electrolyte, the conductivity of the electrolyte changes, which further reduces the potential. For hydrogen oxygen fuel cells with concentrated solutions of KOH or NaOH, this effect is negligible. In an ion exchange membrane cell this may not always be true.

Finally, remember that electrodes have a finite resistance that cause a drop in voltage that must also be subtracted from the open circuit voltage to determine the operating voltage.

Space Shuttle Fuel Cells

Let us look briefly at a real fuel cell, the Space Shuttle's fuel cells. They were hydrogen-oxygen fuel cells. The reactants were stored in supercritical, double walled, thermally insulated, spherical tanks. There was a vacuum between the inner and outer walls, much like a thermos bottle. Each hydrogen tank held 41.6 kg (91.7 lbs) of hydrogen at -251 degrees C (-483.8 F), while each oxygen tank held 353.4 kg (779.11 lbs) of oxygen at -176 degrees C (-348.8 F). (These tanks also carried 22.6 kg (49.8 lbs) for the orbiter Environmental Control and Life Support System [ECLSS] to supply breathing oxygen). The fuel cell was fed by one set of each tank, but more sets could be added for longer duration missions. This was called an EDO, Extended Duration Orbiter. The first time it flew was in Columbia for Spacelab USML-1, a 13 day mission. It last flew on STS-107, the Space Shuttle Columbia disaster, and the tanks were destroyed on re-entry. The Shuttle's tank sets were located in the mid fuselage under the payload bay area. The hydrogen was delivered to the fuel cell at 1724 kPa (250 lbs/in²), and the oxygen at 6205 kPa (900 lbs/in²). There were heaters in the tanks. As the reactants were depleted, the heaters add heat to maintain a constant pressure.

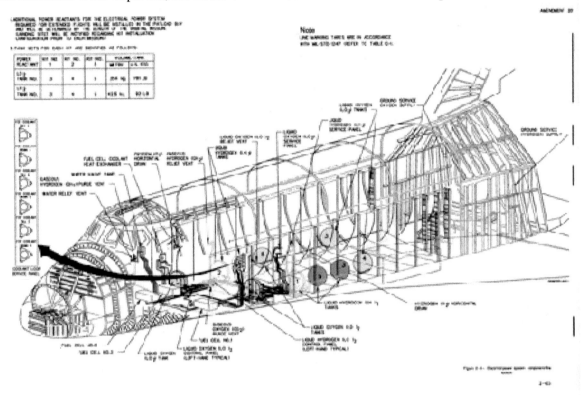

Figure 2.4 Shuttle showing location of fuel cells and reactant tanks (Courtesy NASA)

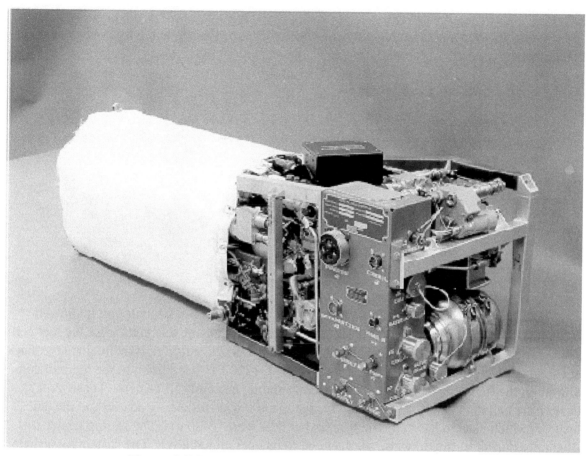

Figure 2.5 Space Shuttle Fuel Cell (Courtesy NASA)

The fuel cells were 35.56 cm high, 38.1 cm wide, and 101.6 cm long (14" high, 15" wide, and 40" long). Each had a mass of 115.66 kg (255 pounds). Each fuel cell had 96 cells in three sub-stacks of 32 cells each. The 32 cells were connected in series, the three sub-stacks were connected in parallel. A single fuel cells produced 10 kW maximum continuous power, 12 kW continuously with one or more fuel cells failed, 16 kW maximum at 26.5 Vdc (emergency, 10 minutes), up to 12 kW at 27.5 Vdc (normal), and 2 kW at 32.5 Vdc. The voltage was regulated to between 27.5 to 32.5 Vdc. (This is because airplanes use 28 volts as their normal voltage.) The orbiter had an average power consumption of 14 kW. The heat generated by the reaction was directed to fuel cell heat exchangers, where excess heat was rejected to the freon coolant loops. The water produced was directed to the potable water storage subsystem. There were manifolds that go the length of the cells to distribute hydrogen, oxygen, water, and coolant.

The cells used a potassium hydroxide electrolyte. At the anode the hydrogen chemical reactions is:

$$2H_2 + 4OH^- \longrightarrow 4H_2O + 4e^-$$

At the cathode the oxygen reaction is:

$$O_2 + 2H_2O + 4e^- \longrightarrow 4OH^-$$

The net reaction is:

$$2H_2 + O_2 \longrightarrow 2H_2O$$

The fuel cells were thermally conditioned by means of the coolant loop. It allowed coolant (freon) to flow through the system to warm the cryogenic reactants, cool the fuel cell stack during its operation, and distribute heat during fuel cell starting. It maintained the stack at a load dependent operating temperature of 93.3 C (200 F). In addition to the thermal conditioning, the fuel cells had internal startup, sustaining, and end cell heaters. These three heaters are called the 1800 Watt startup heater, the 900 watt sustaining heater, and the end heaters (which used to have electric heaters but later were passive). The startup heater was used only during startup to warm the fuel cell to its operational level. The sustaining heaters normally were used during periods of low power to maintain the cell at its operational temperature. The end cell heaters kept the end cells at the same temperature as the cells in the middle of the stack.

When the hydrogen and oxygen entered the fuel cells, they went through a preheater, which warmed them from cryogenic temperatures to 4.44 C (40F) or greater. Then they went through a 6 micron filter and a two-stage gas regulator module. The first stage reduced the pressure of both to 689.47 to 827.37 kPa (100 to 120 psia). The second stage reduced the oxygen to 413.68 to 434.37 kPa (60 to 63 psia) and made the hydrogen pressure at 31.03 to 41.37 kPa (4.5 to 6 psia) differential below the oxygen pressure.

When the hydrogen leaves the dual regulator, it mixed with the hydrogen-water vapor exhaust from the stack. This saturated gas mixture was routed through a condenser, which condensed the water vapor into a liquid. This was separated from the hydrogen-water mixture by the hydrogen pump/water separator. The separator used centrifugal force to remove the liquid and then pressure fed it to the potable water tanks in the lower deck of the crew cabin. The hydrogen pump used 3 phase AC to circulate the hydrogen gas back into the fuel cell, where some of it was used to make power. The rest of it removed the water vapor formed at the hydrogen electrode. The oxygen left the dual gas regulator and went into the fuel cell stack, where it was consumed.

Before launch the fuel cells were on and sharing the load. It T minus 9 minutes, more load was added to the fuel cells. At T minus 50 seconds the GLS (Ground Launch Sequencer) turned off the GSE (ground support equipment) that powered the shuttle and the cells took over the load. Each cell was serviced between flights and were used for 2000 hours, then replaced.

Regenerative Fuel Cells

Regenerative Fuel Cells are simply ones that use electricity to convert the water back to hydrogen and oxygen. You can do this either externally or internally.

Externally would use photovoltaic cells, waste heat from a nuclear power source, or other supplies of electricity to take the water and, using electrolysis, convert the water to hydrogen and oxygen. These are then pumped back to their storage tanks.

Internally uses something called a Unitized Regenerative Fuel Cell (URFC). In this the cell is modified so that putting electricity into the electrodes converts the water back into hydrogen and oxygen. This makes it much lighter than an external system--no external electrodes, pumps, and piping. In the mid 90's, a team at Lawrence Livermore National Laboratory developed a 50 Watt prototype single cell URFC for the Helios aircraft project.

Why use them? If you have an experiment that has a high startup power, such as a superconducting magnet, it may save you mass by using a regenerative fuel cell for peak loads. Once the magnet is up, the solar arrays can supply the continuous power for the magnet plus producing the hydrogen and oxygen.

Efficiency of Fuel Cells

Typically they are in the range of 60-70% for these alkali cells. The Department of Energy says up to 60%. So it depends on what source you read.

Problems

1. A fuel cell consists of 6 cells hooked in series as shown.

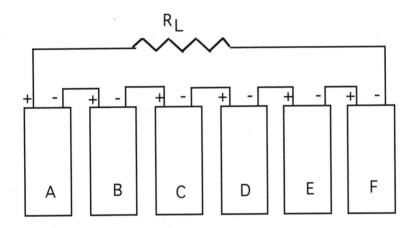

Each cell is hooked to a voltmeter and the following voltages are recorded. All Voltages below are Vac.

$V_A = 0.845V$ $V_B = 0.871V$ $V_C = 0.884V$ $V_D = 0.91V$ $V_E = 0.91V$ $V_F = 0.923V$

When hooked to a load, R_L, the cells output 11.53 amps for 16,736 seconds. $\Delta H = -60$ Kcal/mole, $V = 1.3$ V, $V^0 = 1.42$ V, 1 Kcal = 4184 Joules.

For the System find:

a) η_{ac} of each cell, then of the system of all 6 cells

b) η_V of the whole system

c) n (the number of electrons) of the whole system

Part 2. We now have another fuel cell which consists of 3 of the above, in other words, 18 cells hooked in series with the voltages exactly as above. Therefore, put three of the above together.

$V_A = 0.845V$ $V_B = 0.871V$ $V_C = 0.884V$ $V_D = 0.91V$ $V_E = 0.91V$ $V_F = 0.923V$

$V_G = 0.845V$ $V_H = 0.871V$ $V_I = 0.884V$ $V_J = 0.91V$ $V_K = 0.91V$ $V_L = 0.923V$

$V_M = 0.845V$ $V_N = 0.871V$ $V_O = 0.884V$ $V_P = 0.91V$ $V_Q = 0.91V$ $V_R = 0.923V$

Again, for the system what is

d) η_{ac}

Hint: If we put 42 of the same cells together, what would you think d) would be?? Don't answer this, just think about it.

2. We have the Space Shuttle fuel cells.

The fuel is pure hydrogen and the oxidizer is pure oxygen.

a) Using the Nernst equation what is V?

b) The voltage efficiency, η_v is 70%, what is Vac?

Chapter 3, Semiconductors

The best way to define a semiconductor is to define a conductor and insulator. Conductors allow electrons to pass through them very easily. Typical resistivities are 10^{-6} ohm-cm. Insulators are the exact opposite, they resist the passage of electrons very well. Typical values of resistivity are 10^{12} ohm-cm. Semiconductors aren't great conductors or insulators; 10^{-3} to 10^6 ohm-cm. Semiconductors also may have a negative temperature coefficient of resistance or the ability to produce voltage or decrease resistance when exposed to light. These sorts of things can only be explained by using quantum physics.

Quantum Physics

There are four basics of quantum physics.

(1) Wave particle duality. Sometimes the best way to analyze an electron is to think of it as a particle, and sometimes it is best to look at it as a wave.
(2) The Pauli exclusion Principle. This states that in any atom no two electrons can have the same four quantum numbers.
(3) The Heisenburg uncertainty principle. This states that the exact position and momentum of a particle can never be known simultaneously.
(4) The mathematics of probability assume that all similar particles are indistinguishable.

Another thing to note, the universe is lazy. Matter at room temperature when left to itself goes to the state of least potential energy (PE). This potential energy is not just gravitational PE, and is sometimes called "free energy."

Figure 3.1 shows an experiment that quantum mechanics can explain, but normal physics cannot.

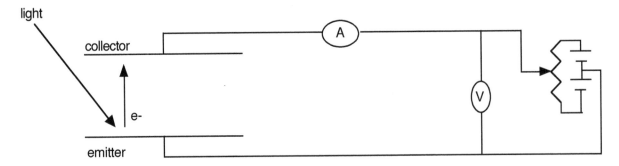

Figure 3.1 Quantum experiment

Light falls onto the emitter and electrons are emitted. The battery supplies a positive charge and the electrons go to the collector, the ammeter A registers a current. By adjusting the voltage you can stop the electrons from reaching the collector and no current flows. This potential is called the "stopping voltage" Vq. Now, as the frequency of the light is increased (shorter wavelengths) the stopping voltage is higher. The minimum frequency that will expel an electron is called the "threshold frequency." No matter how much light is shined on the material, no electrons will be given off unless the light is above that frequency. There is a linear relationship between the frequency of the light and the stopping potential (figure 3.2).

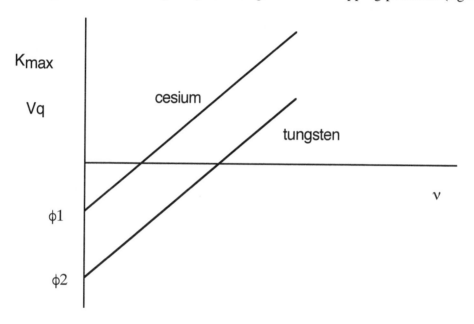

Figure 3.2 Max kinetic energy vs. frequency

$$eV_q = K_{max} = a\nu + b \qquad\qquad 3\text{-}1$$

Where a is the slope and b is a constant. a is independent of the material but b is not.

The wave theory explains why an increase in light intensity increases the amount of current, but fails to explain the threshold frequency. A man by the name of Einstein won the Nobel prize for applying quantum theory to explain this. Quantum theory states that:

(1) Energy is not infinitely divisible, it comes in discrete packets called a quantum. A quantum of light is called a photon.
(2) An electron either gains a quantum of energy or not. The energy is proportional to the frequency of the light:

$$E = h\nu = ch/\lambda \qquad\qquad 3\text{-}2$$

Where h is Planck's constant (6.62×10^{-27} erg-sec), ν is the frequency and λ is the wavelength. The speed of light, c, is related to those two as follows: $c = \nu\lambda$.

(3) The kinetic energy of the electron that absorbs E amount of energy is:

$$K_{(max)} = h\nu - e\phi \qquad\qquad 3\text{-}3$$

Where e is the charge of the electron and ϕ (in volts) is the "work function" of the metal. $e\phi$ is in units of Joules (units of energy), the work function is also given in electron volts (ev) {which is also in units of energy}. The work function is defined as the work required to move an electron from the Fermi level to infinity.

Figure 3.3 is the Electromagnetic Spectrum. As the frequency increases (shorter wavelength) the energy increases. Note how small a part of this is the visible light spectrum.

The Electromagnetic Spectrum

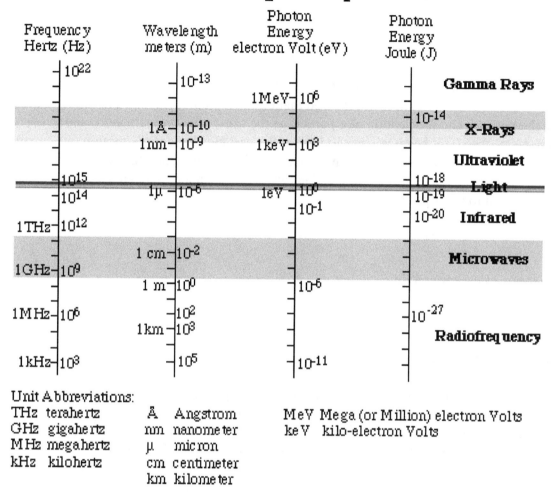

Figure 3.3 Electromagnetic Spectrum (Courtesy NASA)

A. H. Compton, in 1923, did an experiment that helped rectify the differences between the wave and particle theories of light. When monochromatic X-ray waves are beamed at a crystal, classical electromagnetic theory would state that the waves would be scattered, but would keep their frequency. However, he found that in actuality they scatter with some of the waves decreasing in frequency (increased wavelength). This can only be explained if one assumes that the X-rays are particles with energy $h\nu$ and momentum p'. The scattering is then an elastic collision, like a billiard game. The electron recoils, taking some of the energy of the incident photon. The scattered photon has a lower frequency. The photon has momentum:

$$p' = h\nu/c \qquad \qquad 3\text{-}4$$

Louis de Broglie, in 1924, suggested that electrons must be waves also, with wavelengths of:

$$\lambda = h / p' = h / (mu) \qquad\qquad 3\text{-}5$$

Where m is mass and u is velocity

Werner Heisenberg, in 1927, showed that on the atomic level, one cannot determine both the position and momentum of a particle at the same time (you always could in classical mechanics). The light beam you use to look at the particle changes the motion of the particle, for the reasons given above. This is the so called "Heisenberg Uncertainty Principle" which leads to physics students writing on the blackboard, "Heisenberg may have been here." In terms of equations his principle is:

$$(\Delta x)(\Delta p') \geq h / (4\pi) \qquad\qquad 3\text{-}6$$

or in terms of energy and time

$$(\Delta E)(\Delta t) \geq h / (4\pi) \qquad\qquad 3\text{-}7$$

The Schrodinger Equation

Schrodinger, in 1926, came up with a unifying equation that tied all of the new quantum principles together. The equation is a wave equation (it must be to explain the wave nature of the electron diffraction experiments of Compton) and is similar to the equations that define vibrations. We will not solve his equations, merely state the results.

The solution to Schrodinger's equation for the hydrogen atom (which is in three dimensions) yields 3 constants called quantum numbers (that take integer values). The principle quantum number, n, can have any integer value from 1 to infinity. The azimuthal quantum number, l (that's a small L {el}) can have integer values from 0 to n-1 (n - one) and it determines the electrons orbital angular momentum. An electron moving is a moving charged particle and has a magnetic field associated with it. When it interacts with an external magnetic field the orbital can orient itself in 2l+ 1 (2 el + one) different ways. The magnetic quantum number, m_l (m sub el) can have integer values from -l to +l (-el to el), and determines the orientation of the angular momentum vector. Finally, the charge of the electron is assumed to spin, and the spin quantum number, m_s, takes on values of +1/2 or - 1/2 (that's one-half not el/2). The spin quantum number is also called spin up or spin down. Thus the Pauli exclusion principle shows that for two electrons in the same orbit, the first three quantum numbers are the same, and the last number must be different.

The Fermi Level

Enrico Fermi showed that probability of a particular state being occupied is given by the Fermi function:

$$f(E) = \frac{1}{e^{[(E-E_f)/kT]} + 1}$$

3-8

E is the energy of an allowed state as given by Schrodinger's equation for an atom, E_f is the Fermi energy, k is Boltzmann's constant, 1.38049×10^{-23} J/K, and T is the absolute temperature in Kelvin.

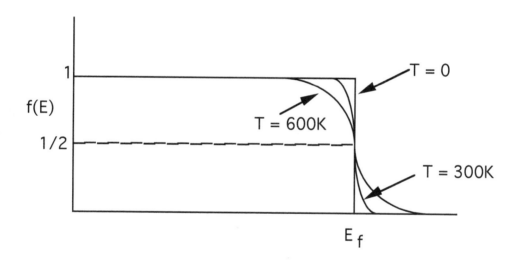

Figure 3.4 Fermi equation

Looking at figure 3.4 with T = 0 we see that for $E < E_f$, all quantum states are occupied. For $E > E_f$, no states are occupied. As T increases the curves change as shown. The Fermi level is therefore defined as the energy at which the probability of a state being filled is 1/2. The Fermi level also is the thermodynamic free energy per electron and is always continuous across the interface of two conductors in equilibrium. We will use this latter later.

Energy Bands

The state of an electron is determined by the quantum numbers n, l, m_l , and m_s . These numbers determine the wave function (the dependent variable of Schrodinger's equation) and the energy. Take two electrons of state (4,1,-1,1/2) and 4,1,1,1/2). They have two different wave functions but have the same energy, since the energy is only a function of the square of the first three numbers. Figure 3.5 shows how we will represent energy levels.

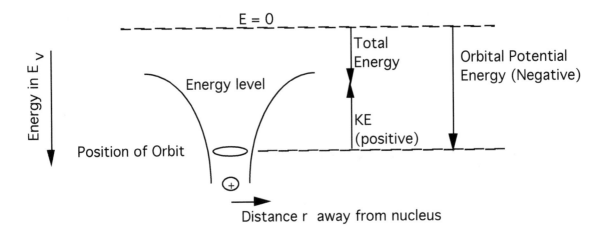

Figure 3.5 Energy levels

The electron is, by ancient conventions, assumed to have zero energy at infinity. As we bring the electron closer to the nucleus, the system does work since the electric charges are opposite. This negative energy is plotted downward. The total energy is the sum of the potential and kinetic, and this locates the energy level.

Figure 3.6 shows a single sodium atom with the location of the electrons in each energy level. Now, when we put a whole bunch of atoms together to form a crystal, we get figure 3.7. Remember the Pauli exclusion principle? Well, now that we have a crystal, the atoms are part of this system and we cannot have more than two in the same energy state. The electrons belong to the crystal, not the individual atoms. The energy levels are now smeared into bands so that there are as many very closely spaced energy levels per band as there are the number of atoms in the crystal.

The highest occupied band is called the valence band. The next allowable band is called the conduction band. The difference in energy levels determines whether the substance is a conductor, insulator, or semiconductor (Figure 3.8).

A conductor requires little energy for an electron to jump into the conduction band and produce current. An insulator has such a wide gap that, under normal circumstances, it cannot accept energy from an applied field. In semiconductors the gap is much lower. At room temperature a few electrons can go into the conduction band. Putting the electrons there is called thermal or optical excitation. As the temperature or frequency is raised, more electrons go up. Not only can these electrons move by reacting to an applied field, but the spaces (called "holes") left by the electrons vacating their old orbits can also move. How do you move a "hole?" You dig a new one. Other electrons move to that site, creating a new hole.

E = 0

Energy in E $_v$

3p (n=3, l=1)
3s (n=3, l=0) Valence level orbit
2p (n=2, l=1)

Position of Orbit

2s (n=2, l=0)
1s (n=1, l=0)

+11

potential energy
curve for electrons

Distance r away from nucleus

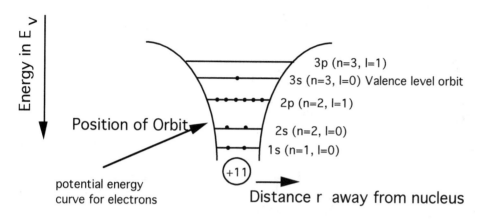

Figure 3.6 Sodium atom

E = 0

Energy bands

potential energy
curve for electrons

+11   +11   +11

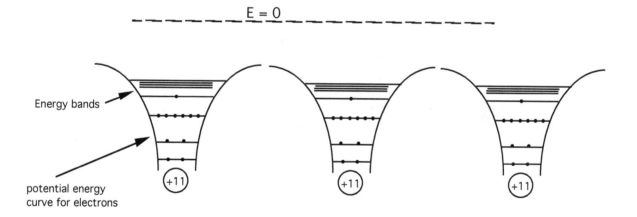

Figure 3.7 Many Sodium atoms

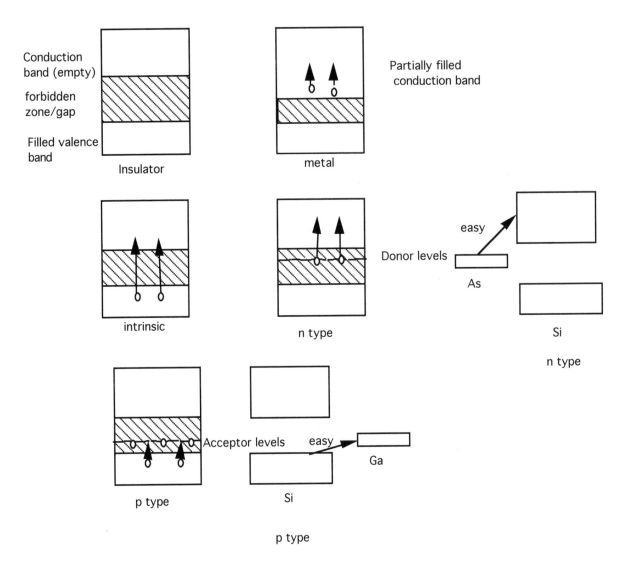

Figure 3.8 Conductors, Insulators, and Semiconductors

Intrinsic Semiconductors

When the only current flowing is caused by electrons that have come up in this fashion (electrons jump up to the conduction band from the valence band) the semiconductor is called intrinsic (figure 3.9).

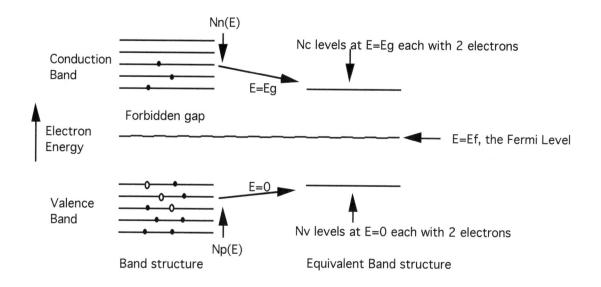

Figure 3.9 Intrinsic Semiconductor

The energy gap from the top of the valence band to the bottom of the conduction band is E_g. The zero level is chosen arbitrarily to be at the top of the valence band. The density of the electron energy levels, $N_n(E)$ is measured up from the bottom of the conduction band. The density of hole energy levels, $N_p(E)$ is measured down from the top of the valence band. For intrinsic semiconductors:

(1) At temperatures above 0 K, there will be a few electrons in the conduction band, and holes in the valence band.

(2) The numbers increase with T.

(3) The numbers of holes and electrons are equal.

(4) The Fermi level is somewhere in the forbidden energy gap.

When we deal with the motions of charge carriers (electrons and holes) in solids caused by external electric fields, we can't apply F=ma, because the electrostatic fields of each atom also cause an effect on the carrier. So, we use something called the effective mass parameter, m*. Because of quantum effects, this may vary from 1/100 to >1 times the electron mass! One advantage of this is that it allows us to replace all the finely spaced energy levels by an equivalent number of energy levels all with energy $E = E_g$. This number is N_c:

$$N_c = (2\pi m^* kT/h^2)^{3/2} \quad \text{per unit volume} \qquad\qquad 3\text{-}9$$

The subscript c refers to the conduction band. You can do the same thing for the holes with a v subscript for the valence band.

If we make the assumption m = m* , which is only good because other errors introduced by changes in temperature, impurity content, etc. are so large, we get:

$$N_c = N_v = (2\pi mkT / h^2)^{3/2} \quad \text{per unit volume} \qquad\qquad 3\text{-}10$$

Taking all the temperature independent stuff out of that and computing its value we get:

$$A = 2.35 \times 10^{15} \text{ cm}^{-3} \text{ K}^{-3/2}$$

Therefore,

$$N_c = N_v = AT^{3/2} \text{ (cm}^{-3}) \qquad\qquad 3\text{-}11$$

The number of electrons per unit volume in the conduction band, n, is the number of
equivalent levels, N_c , times the probability of filling each level (Fermi eq.) times the number
of electrons each level can hold (2). Therefore,

$$n = \frac{2N_c}{e^{[(E_g - E_f)/kT]} + 1} \text{ cm}^{-3} \qquad\qquad 3\text{-}12$$

The number of holes per unit volume in the valence band, p, is the number of electrons
the band could hold minus the number of electrons that are there.

$$p = 2N_v[1 - \frac{1}{e^{[-E_f/kT]} + 1}] \text{ cm}^{-3} \qquad\qquad 3\text{-}13$$

Since n = p you can set these two equal and solve for E_f .

$$E_f = E_g / 2 \qquad\qquad 3\text{-}14$$

Note that the Fermi level is independent of temperature and is in the middle of the gap.
Putting this into 3-12 we get,

$$n = p = \frac{2AT^{3/2}}{e^{[E_g/2kT]} + 1} \text{ cm}^{-3} \qquad\qquad 3\text{-}15$$

In semiconductors that are useful, E_g must be a few kT wide, or it won't be a semiconductor. If we assume it is about 5-6kT then the 1 is insignificant and,

$$n = p = n_i = 2AT^{3/2} \, e^{[-E_g/2kT]} \; \text{cm}^{-3}$$

3-16

n_i is the concentration of either type in an intrinsic semiconductor.

Extrinsic Semiconductors

So much for pure semiconductors. If we look at figure 3.10a we see said pure material. All have 4 valence electrons. If we add an element that has an extra valence electron, such as arsenic, we get an extra electron that requires very little energy to go into the conduction band. These are called donor elements (Figure 3.10b), Similarly, elements with only 3 valence electrons, such as gallium, are called acceptor elements (3.10c). The resulting materials are called n and p type, respectively.

All atoms are Silicon

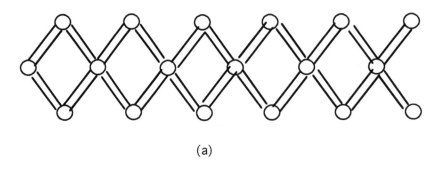

(a)

All atoms are Silicon except one

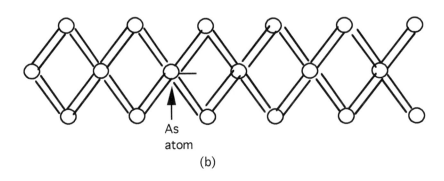

As
atom

(b)

All atoms are Silicon except one

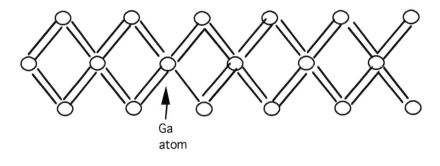

Ga
atom

(c)

Figure 3.10 Pure semiconductor (a), Donor (b), Acceptor (c)

Let's look at a p type (Figure 3.11). The valence band is filled, the conduction band is empty (relatively in both cases) and the energy level of the acceptor states is slightly above the valence band. Electrons can be excited out of the valence band into the acceptor states and create holes in the valence band. There will be more holes in the valence band than there are electrons in the conduction band.

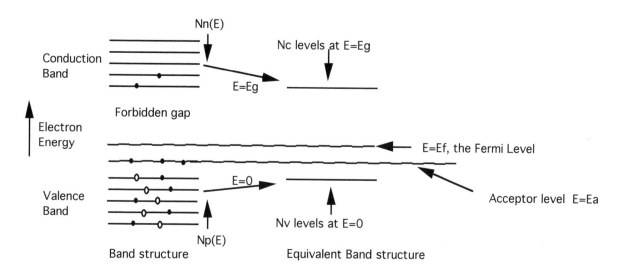

Figure 3.11 p-type semiconductor

To maintain a net charge neutrality, the number of holes per unit volume must equal the number of electrons per unit volume in the conduction band plus the number in the acceptor level.

$$p = 2N_v[1 - \frac{1}{e^{[-E_f/kT]}+1}] \text{ (holes / cm}^{-3} \text{ in the valence band)}$$

$$= \frac{N_a}{e^{[(E_a-E_f)/kT]}+1} \text{ (electrons / cm}^{-3} \text{ in the acceptor level)}$$

$$+ \frac{2N_c}{e^{[(E_g-E_f)/kT]}+1} \text{ (electrons / cm}^{-3} \text{ in the conduction band)}$$

3-17

The acceptor level can only hold one electron per level, the others 2. If the temperature is below the temperature at which the material behaves like an intrinsic semiconductor, then we can ignore the last term and solve for E_f .

$$E_f \approx kT\ln\{\frac{N_v}{N_a} - \frac{1}{2} + \frac{1}{2}\{(1 - \frac{2N_v}{N_a})^2 + (\frac{8N_v}{N_a})e^{[\frac{\varepsilon_a}{kT}]}\}^{\frac{1}{2}}\}$$

3-18

We can reduce this further by assuming the temperature is high enough so that the exponential is close to 1. Then,

$$E_f \approx kT \ln(2AT^{3/2}Na)$$

3-19

Where $AT^{3/2}$ was substituted for N_v in 3-18. Putting 3-19 into 3-17 we get

$$p \approx \frac{2AT^{3/2}N_a}{2AT^{3/2} + N_a e^{[E_a/kT]}}$$

3-20

In extrinsic semiconductors the Fermi level is a function of temperature. For this p - type, as the temperature rises from absolute zero, the Fermi energy level moves from halfway between the top of the valence band and the acceptor level to the midpoint of the energy gap, which happens as the temperature is hot enough to make the material intrinsic.

N type material can be analyzed similarly.

Finally, If you multiply equations 3-12 and 3-13, first making the assumption that the forbidden zone is 5 to 6kT wide and simplifying the Fermi probability expressions like we did before, you get the fact that when you dope an intrinsic semiconductor, as the donors (or acceptors) increase, the holes decrease in concentration. Thus there is a constant in the pn product.

$$pn = n_i^2 \approx (2AT^{3/2})^2\ e^{[-E_g/kT]}$$

3-21

Transport Phenomena

We now have to deal a little bit with how the holes and electrons move through the semiconductor. First we must arm you with some definitions.

The current density, $J = I/A$, is the current divided by the cross sectional area perpendicular to the direction of the current.

The electric field intensity, $E = F/q$, is the force per unit charge. The electrostatic potential difference (the voltage, V) and E are intertwined by the definition of work.

$$V = -\int (F/q)dx = -\int_l^0 Edx = El$$

3-22

Note that the integral goes from l (the letter "el") to 0 because that is how we defined work (from infinity to 0).

Electrical conductivity is the ratio of the current density to the electric field intensity.

$$\sigma = J / E \text{ or } J = \sigma E \qquad\qquad\qquad 3\text{-}23$$

Substituting the current and area for the current density and the voltage for the electric field (from 3-13)

$$I / A = \sigma(V / l) \quad \text{or} \quad V = Il / (\sigma A) \qquad\qquad 3\text{-}24$$

The resistivity is the reciprocal of the conductivity

$\rho = 1/\sigma$. Substituting this into 3-24 we get the resistance

$$R = \rho l / A \qquad\qquad\qquad 3\text{-}25$$

And now we can rewrite 3-24 as the familiar Ohms law

$$V = IR \qquad\qquad\qquad 3\text{-}26$$

The current density is also the number of carriers (per meter cubed) times the electric charge times their speed.

$$J = neu \qquad\qquad\qquad 3\text{-}27$$

Remember that the electron velocity is in the opposite direction of the current density. (One of the reasons it is sometimes easier to think of holes moving!)

The mobility or drift mobility is the indication of how rapidly a carrier will move when an electric field is applied. This is the average drift velocity per unit electric field.

$$\mu = u / E \qquad\qquad\qquad 3\text{-}28$$

Substituting 3-27 into 3-23 we get

$$\sigma = ne\mu \qquad\qquad\qquad 3\text{-}29$$

Example 3-1: Bismuth Telluride

Bi_2Te_3 has a mobility of 540 cm² per volt-sec at room temperature. It also has approximately 5×10^{18} charge carriers per cubic centimeter. Putting these into 3-29

$\sigma = (5 \times 10^{18}\ cm^{-3})(1.602 \times 10^{-19}\ coulombs)(540\ cm^2\ per\ volt\text{-}sec)$
    $= 4.32 \times 10^2\ (ohm\text{-}cm)^{-1}$
$\rho = 1/\sigma = 2.32 \times 10^{-3}$ ohm-cm which is in the semiconductor range.

Armed with all these definitions, let's look briefly at how these carriers move. For example, the electrons are scattered by irregularities in the crystal, so they travel a zigzag path in the opposite direction to the electric field (figure 3.12). Note that the hole travels in an apparent jump if you only look at the beginning and ending of the electron's various meanderings. The average time, τ, between collisions for the electron is the collision time or relaxation time. The distance traveled is called the mean free path. The acceleration is approximately in accordance with Newton's law

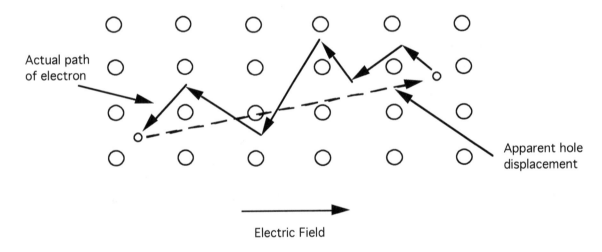

Figure 3.12 Electron motion in semiconductor

$a = F/m^* = eE/m^* = u/\tau$ 3-30

Combining 3-28 into 3-30 we get

$\mu = u/E = e\tau/m^*$ 3-31

and the conductivity is

$\sigma = ne\mu = ne^2\tau/m^*$ 3-32

These assume that all the collision times are the same, which they are not. So we replace them with a summation of all the electrons, n_1 , that have a collision time of t_1 , plus all the electrons , n_2 , with collision times t_2 , etc. The conductivity is

$$\sigma = \sum_j \sigma_j = (e^2 / m^*) \sum_j n_j \tau_j$$

3-33

Multiplying top and bottom by n we get

$$\sigma = ne^2 / m^* \sum_j n_j \tau_j / n$$

3-34

Comparing this with 3-32 we can define an average relaxation time, $< \tau >$

$$< \tau > = \sum_j n_j \tau_j / n$$

And 3-34 and 3-31 become

$$\sigma = ne^2 < \tau > / m^*$$

3-35

$$\mu = e \ < \tau > / m^*$$

3-36

Finally, we have only been considering one charge carrier, a hole or electron. This is great for a metal or semiconductor where only one carrier is dominant, but when both are at work, both contribute to the conductivity so;

$$\sigma = e(n\mu_n + p\mu_p)$$

3-37

Summary

What does it all mean? Well, equation 3-37 can be rewritten using classical statistics. The number of electrons that move to the conduction band via thermal or optical excitation is small versus the total number in the valence band. You get a horrible equation that is a function of temperature, the effective mass parameters of both carriers, both mu's , etc. The net result is as the temperature rises high enough electrons are boosted across the forbidden energy gap and the number of electrons equal the number of holes, and we are in the intrinsic region. In this region the atoms of the crystal are vibrating and colliding with the electrons, scattering them. The mechanical scattering is called lattice, thermal, or acoustic scattering, and sometimes is referred to as electron-phonon collisions. The mean free path decreases as does the mobility. The bottom line, you must keep a semi-conductor cool or else it becomes intrinsic. Intrinsic semi-conductors, can't be made to do anything useful, extrinsic are what several space power systems use.

Problems

1. A certain metal has a work function of 6.8 eV. What is the lowest frequency radiation that can cause the electrons to just be ejected from the surface?

2. With T measured in Kelvin and k the Boltzmann constant, the kinetic energy (KE) of a gas atom is 3/2(kT). What is the corresponding de Broglie wavelength of a hydrogen atom at 30 C? $KE = 1/2mv^2$

3. A scientist has created a material with the following properties.

n = 5.32 x 10^{21} electrons/cm3
μ = 539 cm2 /Volt-sec

He thinks that he can design a better semiconductor using this for an n type material. Can he? Why or why not? (Be brief)

Chapter 4, Thermoelectric Power

History

It is interesting to note that in 1821, the discoverer of this phenomena, Thomas See-beck, didn't figure out exactly what was happening. Instead of current flowing though a wire caused by two different metals being heated, he thought he was getting magnetism caused by the temperature difference. He investigated many materials, some of which would have had an efficiency of 3% in converting heat to electricity. This was comparable to the efficiencies of the steam engines being produced around the world. At that point in history the only machines that produced electric current were very weak electrostatic generators. If he had known the principles of thermoelectric generators, the world today may have a lot of its power generated by thermoelectric instead of steam turbines. Steam didn't turn a generator until 50 years after his research! Extrapolating further, there is the possibility that the vacuum tube would have been invented after the transistor! Think of how that would have changed history. The only use of this property was the use of thermocouples to measure temperature.

In 1834, a French watchmaker and part time physicist, Jean Charles Peltier found that passing an electrical current through a junction of two dissimilar metals would produce heating or cooling at the junction. The heat absorbed or created at the junction is proportional to the electrical current. The proportionality constant is known as the Peltier coefficient.

In 1854 William Thomson (later Lord Kelvin) explained the relationship between the Seebeck and Peltier Effects (known as the Kelvin Relations). The effects are related by The Peltier coefficient is simply the Seebeck coefficient times absolute temperature.

Thermocouples

Figure 4.1 shows a thermocouple circuit.

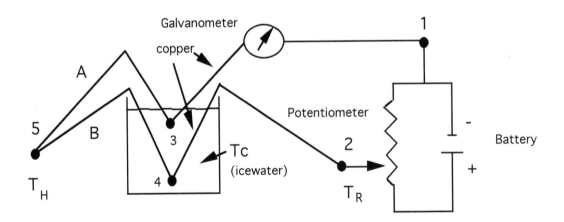

Figure 4.1 Thermocouple

Thermocouples are used to measure temperatures in many situations. Two metals, A and B, are connected together at point 5. At 3 and 4 another metal, let us use copper wire, connects to A and B and is kept at a lower temperature, Tc (usually in an ice bath). These leads go to a potentiometer and battery at room temperature, T_R . When the potentiometer is adjusted so that no current flows in the galvanometer, then the voltage at 1 and 2 is the See-beck Voltage, $V_{A,B}$. The Seebeck voltage is defined to be:

$$\Delta V = \alpha \Delta T \text{ (Current } = 0 \text{ and as } \Delta T \text{ goes to 0)}$$ 4-1

Where α is the Seebeck coefficient. We can now use this to compute the Seebeck voltage in the circuit. We replace the deltas by d and integrate around the circuit.

$$V_3 - V_1 = \int_{T_R}^{T_C} \alpha_{C_u} dT$$

$$V_5 - V_3 = \int_{T_C}^{T_H} \alpha_A dT$$

$$V_4 - V_5 = \int_{T_H}^{T_C} \alpha_B dT$$

$$V_2 - V_4 = \int_{T_C}^{T_R} \alpha_{C_u} dT$$

The right hand side of the first and last equations cancel. The left hand side is V2 - V1, the Seebeck Voltage. Therefore,

$$V_{A,B} = V2 - V1 = \int_{T_C}^{T_H} (\alpha_A - \alpha_B) dT$$

4-2

Thus, the Seebeck voltage is independent of the Seebeck coefficient of the copper wire and of the room temperature .

Figure 4.2 shows a junction of A and B with electric current flowing in it, I.

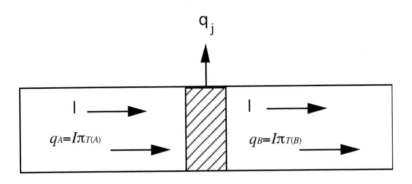

Figure 4.2 Thermocouple junction

The junction is at a constant temperature T. In a normal conductor with electric current flowing in it the conductor heats up due to the resistance of the conductor. This is called Joule heating and is equal to $I^2 R$.

$$P = \text{Heat} = I^2 R \qquad 4\text{-}3$$

If you measure the heat caused in this couple you will find it to be greater or less than the Joule heating! This is called the Peltier effect and is dependent upon A, B, T, and I (magnitude and direction). The amount of heat given or absorbed is given by the Peltier coefficient, π_T (watts per amp). On the left side of the junction, the current produces a heat in of $q_A = I\pi_{T(A)}$ and at the right side the current produces a heat out of $q_B = I\pi_{T(B)}$. If we look at the heat that must be liberated in order to maintain constant T we get, using the first law of thermodynamics:

$$q_j = I^2 R_j + I(\pi_{T(A)} - \pi_{T(B)}) \qquad 4\text{-}4$$

The heat is simply the Joule heating plus the Peltier heating. The "second Kelvin relation" is another method of expressing this mathematically.

$$\pi_{T(A)} - \pi_{T(B)} = T(\alpha_A - \alpha_B) \quad or \quad \pi_{T(A,B)} = T\alpha_{(A,B)} \qquad 4\text{-}5$$

Substituting into 4-4

$$q_j = I^2 R_j + IT(\alpha_A - \alpha_B) \qquad 4\text{-}6$$

Note that the first term is always a loss, but that the second term can be a loss or gain. Reversing the direction of the current reverses the Peltier heating to make it cooling. We could have had a Seebeck refrigerator before the icebox! In fact, you can buy thermoelectric coolers powered by your car today.

Quantum (Solid State) Description of the Thermo Electric Effect

You knew you learned all that quantum from the last chapter for a reason. Figure 4.3 shows a p-type semiconductor connected between two metal contacts. A battery is placed across the contacts to supply a potential difference. The contacts are ohmic, i.e. ohms law applies V=IR.

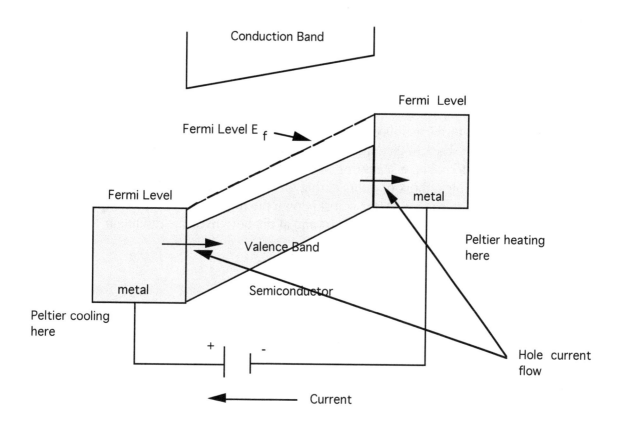

Figure 4.3 p-type connected between two metal contacts with current supplied be a battery

The Peltier coefficient, π_T, shows the amount of cooling or heating that happens at a junction of two different materials over and above the Joule heating that takes place at the contact (eq. 4-4). The Peltier coefficient is the ratio of energy moving into or out of the contact per coulomb of charge going across the contact, its dimensions are voltage. The Peltier heating or cooling can be interpreted as the change in the average kinetic energy of a current carrier when it crosses a contact. Since the metal is in thermal equilibrium with the semiconductor, the Fermi level for the metal is, by definition, the same as the semiconductor's Fermi level. As a hole current flows from left to right in the metal, the holes have, on the average, the Fermi energy of the metal, which is less than the Fermi level of the semiconductor/metal. In order for the hole to cross the junction, it must have an energy of E_f more than the average energy of the metal. You can use transport theory to show that the holes that cross the contact have an average energy of 2kT more than the Fermi energy of the metal. This extra energy comes from the metal. The magnitude of the energy per coulomb (units of volts) is the Peltier coefficient and is the sum of the Fermi energy plus 2kT.

$$\pi_{T(P)} = \frac{(E_f + 2kT)}{e}$$

4-7

The metal lattice provides this energy to the holes, thus cooling it. At the other end the holes impart this energy back to the lattice, causing it to heat. For the system energy is conserved. These results do not take scattering into account, so they are only approximate.

If this sounds strange, think of going skiing on a mountain. The ski lift takes you up to the mountain, giving you energy. After you ski down the mountain, you have expended all your energy into heating the snow at the slope and at the bottom of the mountain. Time for a hot drink!

Analysis of a TEG

Now we will look at a thermoelectric generator. The analysis will not be exact, but the results obtained do not differ very much from an exact analysis. Figure 4.4 shows the device.

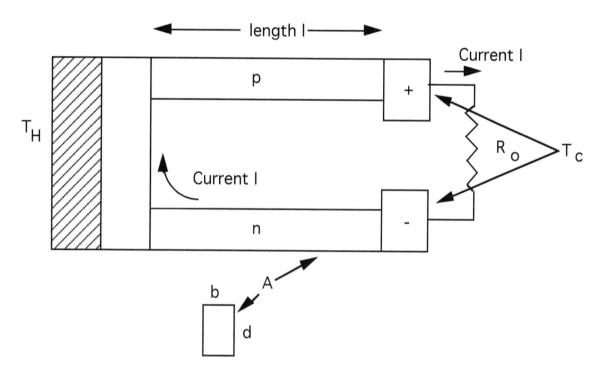

Figure 4.4 Thermoelectric device

There are certain assumptions:

(1) No heat is transferred except through the thermoelectric elements. No heat is transferred laterally from the arms. (this is a good assumption)

(2) The junction resistance is small compared to the resistance of the arms (this we can control by designing properly).

(3) The arms are of constant cross sectional area, bd, b times d.

(4) The resistivity, ρ , thermal conductivity, λ , and Seebeck coefficient, α , are independent of T. (this is a good assumption if T_H and T_c are relatively close in temperature, say a couple of hundred degrees apart)

(5) The thermal contact resistance is small at the source and sink, the electrical insulation is good. (this is a good assumption)

We consider a length dx of one leg (figure 4.5) with unit cross sectional area.

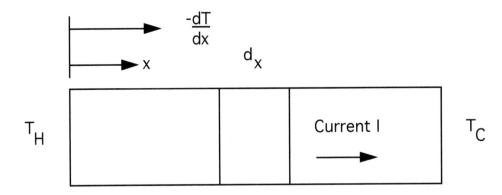

Figure 4.5 One leg of thermoelectric device

The bar has current, I, and temperature gradient, -dT/dx. The heat entering the cross sectional area on the left is:

$$-\lambda dT / dx \qquad\qquad 4\text{-}8$$

Where λ is the thermal conductivity. The heat leaving the right face is:

$$-\lambda\left[\frac{dT}{dx} + \frac{d}{dx}\left(\frac{dT}{dx}\right)dx\right] \qquad\qquad 4\text{-}9$$

The Joule heat in the volume is the current density, J=I/A squared times the resistivity times the length, or:

$$J^2 \rho dx$$

4-10

Since everything is independent of the temperature, heat out must equal heat in plus any heat generated by Joule heating:

$$\lambda \left(\frac{d^2 T}{dx^2} \right) + J^2 \rho = 0$$

4-11

The solution to this may be found by using the boundary conditions:

$T = T_H$ at $x = 0$; and $T = T_C$ at $x = l$ (little el)

$$T = [T_H - (x/l)\Delta T] + [J^2 \rho /(2\lambda)]x(l-x)$$

4-12

Where $\Delta T = T_H - T_C$. The key thing about this is that there are two terms, a linear one, which is the normal temperature gradient in the absence of current, and a parabolic one which is caused by the Joule heating (figure 4.6).

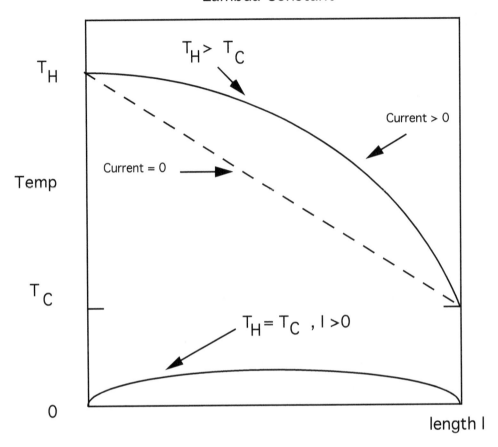

Figure 4.6 Temperature vs. distance in thermoelectric leg

We can divide the heat entering or leaving the TE device into two types, that caused by a temperature gradient at the junction, and that caused by the Peltier effect. The first type can be found by differentiating 4-12 with respect to x and multiplying it by the cross sectional area and the thermal conductivity. For the hot and cold junctions this is:

$$-A\lambda\frac{dT}{dx}\Big|_{x=0} = \gamma\lambda\Delta T - \frac{1}{2}I^2\rho/\gamma \qquad \text{hot junction} \qquad 4\text{-}13$$

$$-A\lambda\frac{dT}{dx}\Big|_{x=l} = \gamma\lambda\Delta T + \frac{1}{2}I^2\rho/\gamma \qquad \text{cold junction} \qquad 4\text{-}14$$

Where $\gamma = A / l$. Note that the heat in is reduced by one half the Joule heating, and the heat out is increased by the same amount.

We can now calculate the Peltier heat term. Since heat will be absorbed between each reservoir and each leg of the generator, the total Peltier heat is defined to be:

$$-(\pi_{T(pr)} - \pi_{T(rn)})I_{pn} = -\pi_{T(pn)}I_{pn}$$

4-15

or, using 4-5

$$-\pi_{T(pn)}I_{pn} = -\alpha T I_{pn}$$

4-16

Where α is the total Seebeck coefficient for the junction $\alpha = |\alpha_n| + |\alpha_p|$. I_{pn} is the current flowing from the p type to the n type arm. The Seebeck coefficients add because the p and n materials act like electromotive forces in series.

Max thermal efficiency of a TEG

The thermal efficiency is defined as the ratio of the electrical power output to the thermal power input to the hot junction.

$$\eta_t = P_o / q_h$$

4-17

The thermal input is 4-13 plus the Peltier effect term of 4-16:

$$q_h = K\Delta T + \alpha T_H I - \frac{1}{2}I^2 R$$

4-18

where K, the thermal conductance, and R, the electrical resistance are defined as

$$K = \lambda_n A_n / l_n + \lambda_p A_p / l_p = \lambda_n \gamma_n + \lambda_p \gamma_p$$

4-19

$$R = \rho_n / \gamma_n + \rho_p / \gamma_p$$

4-20

Note that the current direction is from the n type to the p type (figure 4.4), so the Peltier term has a plus instead of a minus sign. The Peltier effect causes the absorption of heat at the source while half the Joule heat is returned to the source. The power output with a load resistance Ro is:

$$P_o = I^2 R_o (Ohm's\ Law)$$

4-21

The open circuit voltage Voc is $\alpha\Delta T$, so the current I is:

$$I = \frac{\alpha \Delta T}{R + R_0} \qquad\qquad\qquad 4\text{-}22$$

Using 4-17, the thermal efficiency is:

$$\eta_t = \frac{I^2 R_0}{K \Delta T + \alpha T_H I - \frac{1}{2} I^2 R} \qquad\qquad 4\text{-}23$$

Now, sometimes we want to maximize the thermal efficiency. So we use a new variable, m', which is the ratio of the load resistance to the internal resistance.

$$m' = \frac{R_0}{R} \qquad\qquad\qquad 4\text{-}24$$

Rewriting 4-23 using 4-22 and 4-24 and lots of algebra we get

$$\eta_t = \frac{m' \dfrac{\Delta T}{T_H}}{\dfrac{(1 + m')^2 RK}{T_H \alpha^2} + (1 + m') - \dfrac{\Delta T}{2 T_H}} \qquad\qquad 4\text{-}25$$

The significance of this is that, all other things being equal, the smaller RK is, the higher the efficiency. So, we can find RK by using 4-19 and 4-20.

$$RK = \lambda_n \rho_n + \lambda_n \rho_p (\frac{\gamma_n}{\gamma_p}) + \lambda_p \rho_n (\frac{\gamma_p}{\gamma_n}) + \lambda_p \rho_p \qquad\qquad 4\text{-}26$$

You minimize this by taking the derivative with respect to $\dfrac{\gamma_n}{\gamma_p}$ and setting it equal to zero

$$\frac{\gamma_n}{\gamma_p} = (\frac{\rho_n \lambda_p}{\rho_p \lambda_n})^{\frac{1}{2}} \qquad \text{for minimum RK} \qquad\qquad 4\text{-}27$$

and then you get RK minimum by putting this back into 4-26

$$RK_{min} = [(\rho_n \lambda_n)^{\frac{1}{2}} + (\rho_p \lambda_p)^{\frac{1}{2}}]^2 \qquad\qquad 4\text{-}28$$

Now we define a fundamental concept thrown about in all the literature called the figure of merit. It is simply a ratio of the important properties of TEG. The higher the figure of merit, the more efficient the TEG.

$$Z = \alpha^2 / RK$$

If you use the minimum value of RK given in 4-28 you get the maximum value of the figure of merit for any combination of n and p type materials. It is denoted by an asterisk.

$$Z^* = \frac{(|\alpha_n| + |\alpha_p|)^2}{[(\rho_n \lambda_n)^{\frac{1}{2}} + (\rho_p \lambda_p)^{\frac{1}{2}}]^2} \qquad \text{4-29}$$

Now, putting all this together, i.e. 4-27 optimizes the geometry, giving you the smallest RK, thus the best efficiency, and using Z^*, we get:

$$\eta_t = \frac{m' \dfrac{\Delta T}{T_H}}{\dfrac{(1 + m')^2}{Z^* T_H} + (1 + m') - \dfrac{\Delta T}{2 T_H}} \qquad \text{4-30}$$

The value of m', the resistance ratio, that maximizes this is found by taking the derivative and setting equal to zero:

$$m'_{opt} = (1 + Z^* T_{av})^{\frac{1}{2}} \qquad \text{4-31}$$

where $T_{av} = (T_C + T_H)/2$. Putting 4-31 into 4-30 we optimize for both geometry and load resistance and get:

$$\eta_{t(max)} = \frac{(m'_{opt} - 1) \dfrac{\Delta T}{T_H}}{m'_{opt} + \dfrac{T_C}{T_H}} \qquad \text{4-32}$$

So, the thermal efficiency under these conditions is a function of $Z^* T_{av}$ and $\dfrac{T_C}{T_H}$ (figure 4.7).

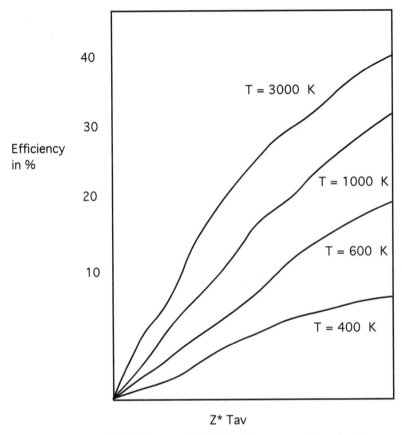

Figure 4.7 Efficiency vs Z* Tav (an approximation)

Max power output of a TEG

Sometimes we wish to maximize the power output. Remember that at maximum power output $R_0 = R$. Taking equation 4-21 and substituting 4-22 for I and rearranging 4-24 for R_0, we get

$$P_O = \frac{(\alpha \Delta T)^2 m'}{(1 + m')^2 R}$$

4-33

Doing derivatives, etc. yields for a maximum

$$\eta_{t(mp)} = \frac{\dfrac{\Delta T}{T_H}}{\dfrac{4}{Z^* T_H} + 2 - \dfrac{\Delta T}{2 T_H}}$$

4-34

74

Practically speaking, when you design a TEG for max power output you try to minimize volume, weight, or TE material. To accomplish this, you must maximize power output per unit of total cross sectional area, P_O / A_{tot} . You also want to keep the length of each element short, but if too short the contact resistance is no longer negligible in comparison to the element resistance (remember our assumptions?) P_O / A_{tot} can be found by setting m' equal to 1 in 4-33 and dividing by the area of the elements, A_n and A_p .

$$\frac{P_O}{A_{tot}} = \frac{(\alpha \Delta T)^2}{4l[(\rho_n / A_n) + (\rho_p / A_p)][A_n + A_p]}$$

4-35

where $A_{tot} = A_n + A_p$. We assume that the lengths of the n and p type material are the same. This is a maximum when the denominator is a minimum, so we take a derivative with respect to A_n / A_p , we get the area ratio that maximizes 4-35.

$$\frac{A_n}{A_p} = [\frac{\rho_n}{\rho_p}]^{\frac{1}{2}}$$

4-36

Look at 4-27, which maximized thermal efficiency. If the lengths were the same we would get

$$\frac{A_n}{A_p} = [\frac{\rho_n \lambda_p}{\rho_p \lambda_n}]^{\frac{1}{2}}$$

4-37

The difference is that the maximum power case does not have any thermal conductivity.

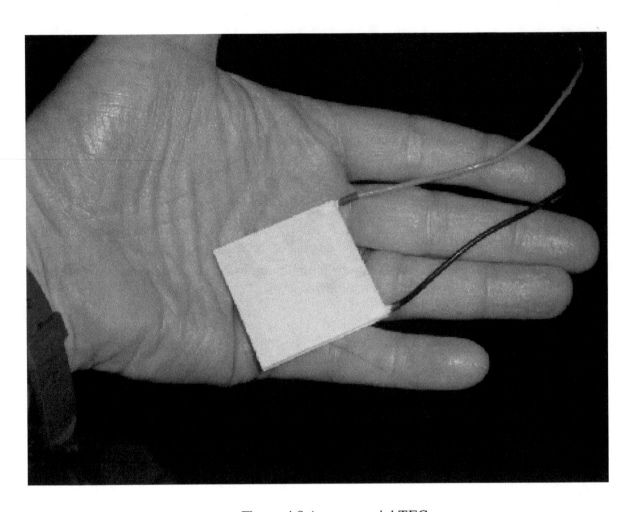

Figure 4.8 A commercial TEG

Real TEGs, especially used in space, need to hook many of the single generators in series and parallel to get realistic voltages and currents (figure 4.9).

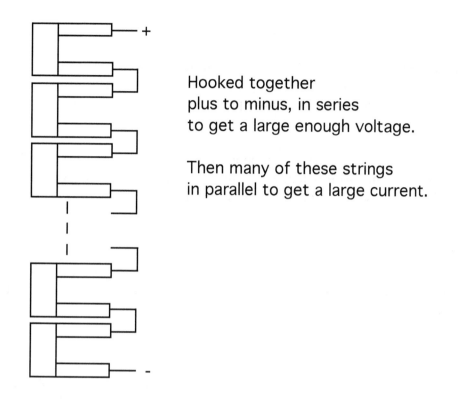

Hooked together
plus to minus, in series
to get a large enough voltage.

Then many of these strings
in parallel to get a large current.

Figure 4.9 String of individual thermoelectric generators

Efficiency of TEGs

In general, the efficiencies are about 9-10%

Problems

1.   Find the current that will make the junction heat transfer be zero for a thermocouple with a Seebeck coefficient of 200 microvolts per K; the temperature of the junction is 600 K with a junction resistance of 0.01 ohm. Draw a picture of this junction indicating the direction of current flow.

2. We are going to design a TEG that will produce 100 watts operating between 50 degrees and 250 degrees C. Assume the efficiency is 7.78%. This TEG, consisting of many of the individual cells hooked in series and parallel, is going to drive a load with a resistance of 8.9 ohms. Find:

a)   The load voltage
b)   The least collector area required. Assume the generator is in earth orbit with incident radiation of 0.135 W/ cm2. Assume that the collector that takes the heat to the junctions has an efficiency of 82%.

3. These are the specifications of a thermoelectric cooler. "cools food to 40 deg F below ambient, draws 4.5 Amps at 12 Volts." Now, I rip the cooler apart to make a TEG. First, the heat that a material will absorb or release is given by

$$q = mc(\Delta T)$$ equation 1

Where m is mass, c is the specific heat in kJ/kg-K, and T is the temperature in K. This heat is in Joules, the heat we have used in class is J/s or watts. We'll take care of that next.

We assume:
m of the device is 0.5kg
c = 0.1 kJ/kg-K
q determined by equation 1 is put in every second.

Warning, the specifications used F not C F = 9/5 C + 32, but delta T is different.

What is the efficiency of the generator?

Chapter 5, Radioisotope Thermoelectric Generators

History

The RTG was invented in 1954 by Mound scientists Ken Jordan and John Birden. They had an Army Signal Corps contract beginning on January 1, 1957, to conduct research on thermocouples and radioactive materials (Polonium-210) suitable for the direct conversion of heat to electrical energy.

The first RTG launched into space by the United States was SNAP 3B in 1961 powered by 96 grams of plutonium-238, on a Navy Transit 4A navigation spacecraft.

Safety

For Radioisotope Thermoelectric Generators (RTG's) the current aerospace safety policy is to provide containment that is not invalidated under any circumstances, including launch accidents (i.e. explosions), the heat of reentry, or the impact on the Earth's surface or water. This was not always the case. The SNAP-3B (Systems for Nuclear Auxiliary Power) and 9A systems were designed to burn up and disperse at high altitude. On April 21st, 1964 a Transit-5BN-3 satellite failed to achieve orbit and the SNAP-9A RTG burned up on reentry as designed. Four months later PuO_2 was detected at an altitude of 32.9 km, this led to the conclusion of total burn up as designed.

On May 18th, 1968 the Nimbus B-1 meteorological satellite with a SNAP-19 on board had an erratic launch vehicle and was blown up at an altitude of 30 km after being launched from the Vandenberg Air Force Base launch site on the southern coast of California. The SNAP-19 was designed to survive a reentry onto land and survive in salt water. The SNAP-19 was recovered from 90 meters of ocean 5 months later. The fuel was re-used in another generator. No detrimental effect were suffered to the fuel capsule despite the treatment of launch abort (explosion), crashing into the sea, and submersion in the ocean for five months.

In April 1970 Apollo 13 had a SNAP-27 on the Lunar Excursion Module (LEM) which reentered the atmosphere at 122 km above the South Pacific. Atmospheric monitoring has proven that it survived reentry and is sitting near the Tonga Trench in 6 to 9 km of water.

Radioisotope Thermoelectric Generators

RTG's have been in use in the space program since 1961. They have powered satellites, Apollo lunar experiments, deep space probes, and some Earth Orbiters. Their basic operation is to use the Radioisotope as a source of heat for a TEG. Plutonium has a half life of 87.75 years. This means that if an RTG starts out with 100 watts thermal of Plutonium, in 87.75 years it will have 50 watts thermal.

The SNAP-3B was the first used in space. It had a mass of 2.1 kg and produced 2.7 watts electric to power the Transit navigation satellite. The most advanced RTG's are in use for the Galileo mission, consisting of 54.4 kg producing 290 watts electric.

A good graphic web page that shows all the missions is at:
http://www.spacesafetymagazine.com/aerospace-engineering/nuclear-propulsion/infographic-atomic-batteries-deep-space-missions/

RTG's have the following advantages:

1) Long life
2) Very reliable
3) Ability to operate in a hostile environment (as in solar flares, high magnetic fields, etc.)
4) Independent of Solar Flux both distance, shading, and angle
5) High power density

RTG's also have the following disadvantages:

1) The high cost of Plutonium 238 (Pu 238)
(For example, the RTG's for Galileo produce 290 Watts electric, mass 11 kg [10.872 kg, to be precise], and cost $8 Million. This is $27,586 per watt electric, $727.27 per gram of Plutonium or 0.02636 Watts per gram!!)

2) Safety--Plutonium is radioactive, although not very.

There are a variety of factors that are considered for selection of a fuel for RTG's.

1) Power density
2) Half-life
3) The cost of the fuel and how easy it is to procure or make
4) External radiations that occur with the decay of the radioisotope (gamma rays and neutrons)
5) Materials science and fuel form inventions that affect potential biological and radiological hazards of accidentally released fuel
6) Negative effects if that fuel is accidentally released into Earth's biosphere

To date all US RTG's used in space have been fueled by Pu 238 because:

1) Alpha emitters are easily shielded (there are a few gamma rays and neutrons emitted, but these are minor problems)
2) They are readily available
3) They have a long half-life
4) Their power density is favorable

The venting of Helium given off by the Plutonium is a small problem. Plutonium emits and alpha particle, which is a Helium nucleus. This nucleus, having no electrons, steals 2 electrons from the surrounding material, making Helium gas.

Other fuels that have been considered are:

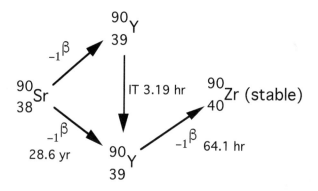

$Sr_{-1}\beta$ max = 0.546 MeV

$Sr_{-1}\beta$ ave = 0.196 MeV

$Y_{-1}\beta$ max = 2.284 MeV

$Y_{-1}\beta$ ave = 0.546 MeV

IT is Internal Transition

Strontium. Disadvantage: low power yield

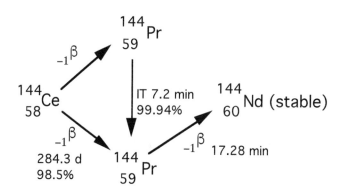

IT is Internal Transition

Cesium. Power density better than Sr

$$^{210}_{84}Po \xrightarrow[138.37 \text{ d}]{\alpha} {}^{206}_{82}Pb \text{ stable} \quad E_\alpha = 5.304\text{MeV (99.999\%)}$$

Polonium. Cheap, but short half-life

$$\,^{242}_{96}\text{Cm} \xrightarrow[163.2\ \text{d}]{\alpha} \,^{238}_{94}\text{Pu} \xrightarrow[87.75\ \text{yr}]{\alpha} \,^{234}_{92}\text{U} \quad 2.445 \times 10^{5}\text{yr}$$

$$\,^{244}_{96}\text{Cm} \xrightarrow[18.11\ \text{yr}]{\alpha} \,^{240}_{94}\text{Pu} \xrightarrow[6569\ \text{yr}]{\alpha} \,^{236}_{92}\text{U} \quad 2.34 \times 10^{5}\text{yr}$$

Curium. Prime contender for replacing $\,^{238}_{94}Pu$

The advances in RTG technology are measured by 3 factors:

1) Increased energy conversion efficiency (Silicon-Germanium [Si-Ge])
2) Higher specific power
3) Longer lifetimes (of total system)

There are two approaches to decreasing the costs of RTG's

1) Increase efficiency of power generation
2) Use less expensive fuel (Cm 244)

However we need additional fuel at BOL to compensate for the shorter half-life of Curium.

Radiation emitted by Cm 244 also requires additional shielding and modification of ground handling equipment (gamma rays also are given off by alpha emissions).

Conversion efficiencies for RTG's has increased due to

1) Material advances (higher Z, figure of merit)
2) Increased operating temperatures

Today a maximum temperature is set by the isotope fuel capability, which is currently 373 degrees K higher than the Thermoelectric material. A higher temperature TE material can operate for longer lifetimes.

We can increase specific power by

1) Developing/using low density thermal insulation
2) Developing low weight structures for support

Table 5.1 compares the half lives, Watts/gm, and $/Watts for various radioisotopes.

Table 5.1 half lives, Watts/gm and price per Watt.

	Po 210	Pu 238	Ce 144	Sr 90	Cm 242
Half-life (years)	0.378	87.75	0.781	28	0.445
Watts/gm (thermal)	140	0.55	25	0.93	120
$/Watt (thermal)	570	3000	15	250	495

All of the images following are Courtesy NASA, except where noted.

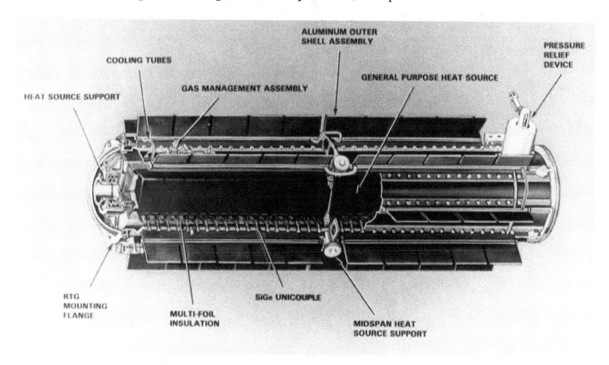

Figure 5.1 General Purpose Heat Source (GPHS)

Two nuclear generators (General Purpose Heat Source, GPHS) power the Galileo spacecraft. Each is about 45 inches long, 16 inches in diameter, and weighs 55 kg (122 pounds). The radioisotopes generate 4300 Watts thermal, and are converted to 300 (or so)

Watts electric. There are 576 thermocouples oriented in 16 circumferential rows each having 36 couples. The output voltage is 30 volts regulated by the electronics. The hot junction temperature is about 1273 K and the cold junction temperature is about 566K.

GPHS-RTG

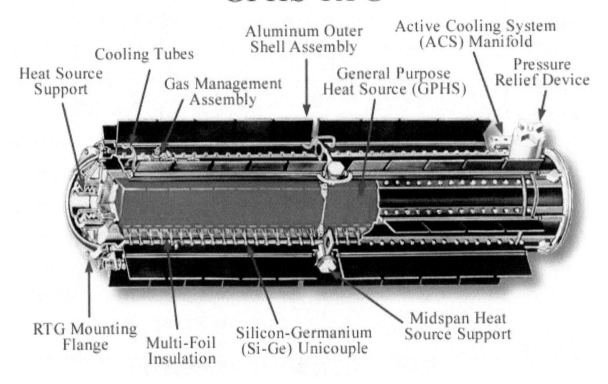

Figure 5.2 General Purpose Heat Source (GPHS) RTG

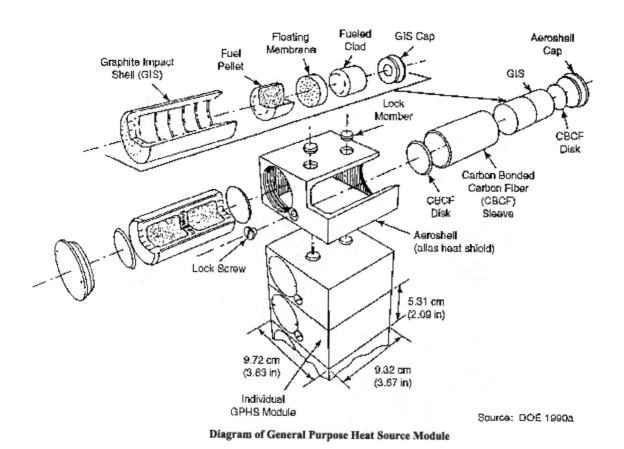

Diagram of General Purpose Heat Source Module

Figure 5.3 GPHS fuel module (Courtesy the Department of Energy (DOE))

Figure 5.4 RTG on the Moon for the Apollo 14 ALSEP experiment is in the foreground.

Figure 5.5 Cassini's RTGs after installation. Note the radiators.

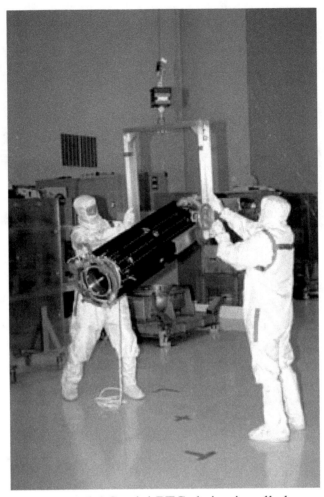

Figure 5.6 Cassini RTGs being installed.

Efficiency of RTGs

In general, the efficiencies are the same as thermo electric generators of about 6-9%. The GPHS has an efficiency of 6.8%.

Problems

1. An RTG is designed using Strontium 90. It is going to produce 100 Watts electric. The Thermoelectric materials are 7.8 percent efficient in converting thermal energy to electricity.

a) How many grams of $_{38}^{90}Sr$ do you need?

b) How much would that many grams cost?

c) If we used $_{84}^{210}Po$ how many grams would you need?

d) How much would that many grams cost?

Chapter 6, Photovoltaic Generators

Definitions.

Photon Flux is the number of photons crossing a unit area at right angles to the light beam per second. If we assume that the energy, E of the photons is approximately the energy of the average frequency, then

$$\Phi = N_{ph}E = N_{ph}h\nu_{av} = N_{ph}h\frac{c}{\lambda_{av}}$$

6-1

N_{ph} is the number of photons impinging on the area (cm squared) per second. Table 6.1 gives the photon flux for various values of m (the optical path length) and w (the number of cm of precipitable water vapor). m = 1/cos z, where z is the angle between the observer and the zenith line and the observer and sun line. For m=w=0 outside the atmosphere N_{ph} = 5.8 x10^{17} per cm^2 per second.

Table 6.1 photon flux versus m.

m	w	location	phi (Φ)	$E_{av\,(eV)}$	N_{ph} (#/sec/cm^2)
0	0	Outside the atmosphere	0.135	1.48	5.8×10^{17}
1	0	Sea Level with the Sun at Zenith	0.106	1.32	5.0×10^{17}
2	0	Sea Level with the Sun at 60 degrees from the zenith	0.088	1.28	4.3×10^{17}
3	0	Sea Level, with the sun at 70.5 degrees from zenith	0.075	1.21	3.9×10^{17}
1	2	~50% Relative Humidity	0.103	1.25	4.8×10^{17}
3	5	Very Extreme Conditions	0.059	1.18	3.2×10^{17}
1	0	Cloudy day with 7000 K black body conditions	0.012	1.44	5.2×10^{17}

Photoconductivity is the increase in conductivity caused when a semiconductor is irradiated with light of the proper frequency. When light hits the semiconductor it produces f hole-electron pairs per cm^3 per second. The increase in hole and electron density is:

$$\Delta n = f\tau_n^* \quad \Delta p = f\tau_p^* \tag{6-2}$$

τ^* is the effective lifetime of the carriers with subscripts n and p. This is NOT the same quantity as the mean relaxation time. Using equation 3-37 from chapter 3

$$\Delta\sigma = e(\Delta n\mu_n + \Delta p\mu_p) = ef(\tau_n^*\mu_n + \tau_p^*\mu_p) \tag{6-3}$$

The photocurrent I_s produced by the illumination is:

$$I_s = eF'\tau^* / T_r \tag{6-4}$$

where F' is the total number of electrons and holes produced each second by the photons being absorbed, and T_r is the transit time. The latter is the time an electron spends going between the two electrodes:

$$T_r = L/<u> \ = L/\mu E \tag{6-5}$$

where <u> is the average drift velocity μ is the drift mobility and L is the distance between the electrodes called the inter-electrode spacing. E is the electric field. Since E = V/L

$$T_r = L^2 / \mu V \tag{6-6}$$

Looking at 6-4 we see that the larger the effective lifetime, the higher the photocurrent. If we assume that the electron lifetime is greater than the hole lifetime, the holes are trapped in a "recombination center" or "trapping center". These centers are a spatially localized energy state created by donor and acceptor impurities, dislocations of atoms, interstitial atoms, etc. Consider a state in an n-type material above the valence band. These states are normally empty, but there is a finite probability that a free electron can drop from the conduction band to this state, releasing a photon to conserve energy. This crystal imperfection causing the state is a trap for electrons. P-type material can have hole traps. "Recombination center" is the name applied to an imperfection that has a high probability of capturing an electron from the conduction band and then losing the electron to the valence band, therefore capturing a hole. Recombination centers are generally located near the center of the forbidden band.

The electrons exist long enough to be moved out by the applied electric field, and the negative electrode injects another electron until the free electron can recombine with the hole. Therefore, the photon appears to produce more than one electron. This gain is denoted by a gain factor, G.

$$G = \tau^* / T_r \tag{6-7}$$

This is an indication of the efficiency of a photoconductor (look at 6-4). It can be increased by decreasing the transit time through the inter-electrode spacing. It can also be increased by increasing the effective lifetime, which is inversely related to the number of sites available for trapping, the effectiveness of the trapping centers, and the drift velocity.

P-n Junctions are the most efficient photo-conversion devices. They can be produced by a variety of ways, and they are simply a layer of p and n material as a single crystal. Since the concentration of electrons is greater on the n side, the electrons diffuse down this concentration gradient into the p side where they combine with holes. (ditto the holes to the n side). In a billionth of a second the process stops, and the n-type region has a plus charge, while the p-type has a negative charge. A contact potential difference, $\Delta\varepsilon$, builds up across the junction to just stop the movement of electrons and holes. The area near the junction now has a space charge or transition region, which is 10^{-4} to 10^{-6} cm wide. See Figure 6.1.

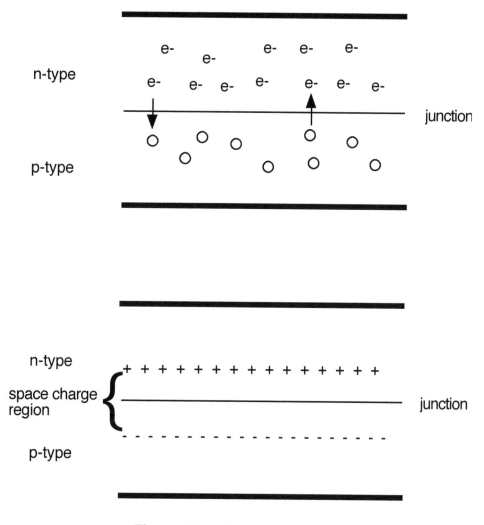

Figure 6.1 p-n junction (n over p)

So, how does a photovoltaic cell work? Figure 6.2 shows how. If a photon hits an electron in the space charge layer, it gains more energy and moves from the valence to the conduction band. The space charge layer pushes the electron towards the n-type, and the hole towards the p-type. If you connect a wire between the two, current flows.

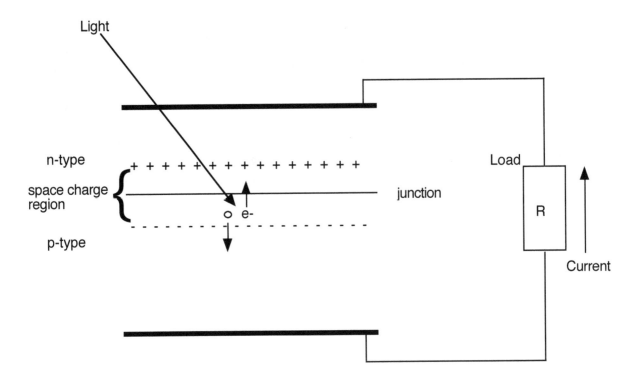

Figure 6.2 Photocell

How efficient is a photovoltaic cell? Well, this will take a while, with many equations, but we'll find out a theoretical maximum efficiency. There are a small number of thermally excited electrons that go into the conduction band, and they go towards the n-type region as well. This gives us a thermal current, I_g, which is directly proportional to the number of thermally excited electrons in the p-type region:

$$n_p \cong A_1 e^{[-(\varepsilon_{g(p)} - \varepsilon_f)/kT]}$$

6-8

Where A_1 is a proportionality constant. In the n-type region, the number of thermally excited electrons in the conduction band is given by:

$$n_n \cong A_1 e^{[-(\varepsilon_{g(n)} - \varepsilon_f)/kT]}$$

6-9

If the latter electrons have enough energy to cross the potential barrier $\Delta\varepsilon = \varepsilon_g(p) - \varepsilon_g(n)$ (the energy gap of the p material minus the energy gap of the n material), they can enter the p-type region and combine with the holes. This produces a recombination current, I_r which is proportional to:

$$A_1 e^{[-(\Delta\varepsilon + \varepsilon_{g(n)} - \varepsilon_f)/kT]} = A_1 e^{[-(\varepsilon_{g(p)} - \varepsilon_{g(n)} + \varepsilon_{g(n)} - \varepsilon_f)/kT]} = n_p \cong A_1 e^{[-(\varepsilon_{g(p)} - \varepsilon_f)/kT]}$$

6-10

And, as you probably guessed, at thermal equilibrium these are the same, $I_r = I_g$.

If we forward bias the junction by hooking a battery so that the + is on the p side and the - is on the n side, the applied voltage reduces the energy barrier to the thermally excited electrons on the n side. So, I_r increases, but I_g stays the same--we've done nothing to change the number of thermally excited electrons on the p side. I_r is increased by the Boltzman distribution factor $e^{[eV/kT]}$. Since I_g does not change and the two are equal at equilibrium:

$$I_r = I_g e^{[eV/kT]}$$

6-11

The net electron current is the difference between the two currents, or

$$I_r - I_g = I_g \{ e^{[eV/kT]} - 1 \}$$

6-12

The hole current behaves similarly, and the two currents add because they are oppositely charged moving in opposite directions. The total current (the junction current) is

$$I_j = I_0 \{ e^{[eV/kT]} - 1 \}$$

6-13

Where I_0 is called the saturation or dark current. This equation is called the rectifier equation. When light hits the p-n junction, current I flows in the load, the magnitude being the difference between the current that would flow if the junction is short circuited, I_s, and the current that flows across the junction, I_j.

$$I = I_s - I_j$$

6-14

$$I = I_s - I_0 \{ e^{[eV/kT]} - 1 \}$$

6-15

Now, we want to make this device efficient and to discover what properties make it more efficient. PVs are usually analyzed in terms of current densities, because it is easier to compare different ones that way. Current density, J, is in units of Amps/cm^2. So, all the I's become J's in 6-14 and 6-15 and we get 6-16 and 6-17,

$$J = J_s - J_j$$

6-16

$$J = J_s - J_0 \{ e^{[eV/kT]} - 1 \}$$

6-17

The maximum voltage would occur when the cell is open circuit, i.e. J=0.

$$V_{oc} = (kT/e)\ln(J_s/J_0 + 1)$$

6-18

To find the voltage that will produce the maximum power density we figure the power output of the PV device:

$$P = JV = (J_s - J_0\{e^{[eV/kT]} - 1\})V$$

6-19

Taking a derivative, setting it to zero, etc. yields

$$e^{[eVmp/kT]}[1 + eVmp/kT] = 1 + J_s/J_0 = e^{[eVoc/kT]}$$

6-20

The current density that maximizes the power comes from combining 6-17 with 6-20

$$J_{mp} = \frac{[eV_{mp}/kT]J_s}{1 + eV_{mp}/kT}[1 + \frac{J_0}{J_s}]$$

6-21

The maximum power density is $P_{max} = J_{mp} V_{mp}$

$$P_{max} = \frac{[eV_{mp}/kT]V_{mp}J_s}{1 + eV_{mp}/kT}[1 + \frac{J_0}{J_s}]$$

6-22

Since the dark current density is five or more magnitudes smaller than the short circuit density the last term is essentially 1. The efficiency is P out divided by P in. The power in is the number of photons, N_{ph} times the average energy of each photon, ε_{av}. Therefore,

$$\eta_{max} \approx \frac{[eV_{mp}/kT]V_{mp}J_s}{[1 + eV_{mp}/kT]N_{ph}\varepsilon_{av}}$$

6-23

This will be simplified further as we go along. One thing to note, the number of photons with energy greater than ε_g decreases as ε_g increases, while the ratio J_s/J_0 and consequently V_{mp} increases with ε_g; obviously η_{max} will pass thru a maximum as a function of ε_g.

We have assumed that the shunt resistance R_{sh} is much greater than the load resistance R_0, and the internal resistance R_s is much less than R_0. The first is easy to do, the latter more difficult. The larger R_s the more power is dissipated at the junction and NOT delivered to the load. Figure 6.3.

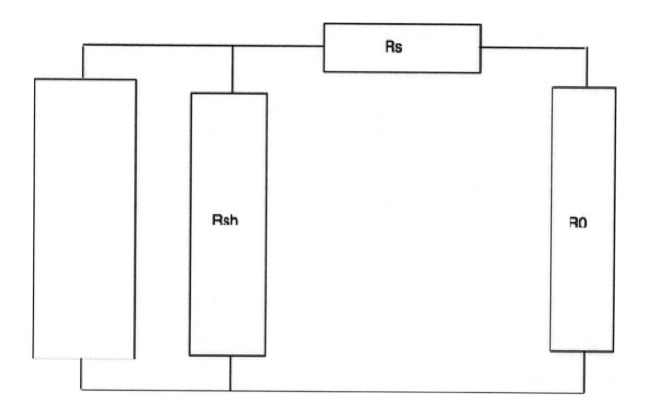

Figure 6.3 Resistances

Properties desired in a semiconductor for cell use

So much for macroscopic parameters such as dark current density, short circuit current density, and voltage at max power output. These tell us how to design a cell, but nothing about which semiconductor is better. We now look at microscopic stuff.

To achieve high conversion efficiency you must produce electron-hole pairs within a very short distance of the p-n junction, otherwise they just recombine without contributing to the cell's output. The average distance a carrier diffuses before recombining is called the diffusion length. They diffuse because of the concentration gradient. The carriers have finite lifetime τ^* and the average distance before recombination, called the diffusion length L, is related to the diffusion constant D by

$$L = (D\tau^*)^{1/2} \qquad\qquad 6\text{-}24$$

To interpret D physically, consider a semiconductor that has a concentration gradient of carriers that varies from being small on the left side of a block of semiconductor to being large on the right side. The carriers motion is response to the gradient is analogous to the mixing of two different fluids. The electric current density that the gradient will produce in the x direction is

$$J_D = -eD\frac{dn}{dx} \qquad\qquad 6\text{-}25$$

The minus sign occurs because diffusion occurs toward the region of decreasing concentration. If this diffusion produces an electric field E, steady state conditions are reached when the conduction current $J_{CD} = \sigma E$ is equal to the diffusion current J_D.

$$ne\mu E = -eD\frac{dn}{dx} \qquad\qquad 6\text{-}26$$

The field will cause a potential energy difference of magnitude eEx to exist over a distance x. Thus, if the carrier concentration is low enough so that we may use Boltzmann statistics to describe the energy distribution of the carriers, then the concentration gradient as a function of x is

$$n = Ce^{[-eEx/kT]} \qquad\qquad 6\text{-}27$$

Where C is a constant. Combining those two equations

$$D = (kT/e)\mu \qquad\qquad 6\text{-}28$$

Which is known as the Einstein relation. This gives us a way to find D because μ can be found from Hall effect experiments.

Back to the series resistance. Since the carriers have to be produced near the junction (10^{-6} cm $< L < 10^{-2}$ cm), the junction is located quite near the top of the cell. In the case of n on p-type the n type layer is 10^{-4} cm thick. This thin layer carrying all the load current is the origin of R_s. Therefore the thickness is a compromise of the value of R_s and the collection efficiency of the junction.

Eq. 6-23 shows that the efficiency is linear with J_s. This is proportional to the efficiency with which the carriers generated in the bulk are collected and delivered to the external circuit, the fraction of the incoming photons absorbed in the effective volume, the fraction of the radiation not transmitted completely through the junction, and the number of photons per second per unit area of p-n junction whose energy is great enough to generate electron-hole pairs in the semiconductor.

$$J_s = \eta_{co}(1-r)[1-e^{(-\alpha l)}]en_{ph}$$

6-29

Where η_{co} is the collection efficiency, r is the reflection coefficient, $e^{(-\alpha l)}$ (e as in the number 2.718281) is the fraction of the radiation transmitted, α is the absorption coefficient, l (little el) is the thickness of the absorbing semiconductor, e is the charge of the electron, and n_{ph} is the number of photons per second per unit area whose energy is great enough to generate hole-electron pairs. Note that in $e^{(-\alpha l)}$ the e is the base e, that is 2.718281 etc. and is not the electric charge.

The mathematical description of light being absorbed in a solid is very similar to the attenuation of gamma or x rays. Photons are removed either by collisions that produce excited states (of the electrons) or free electrons and holes. If Φ_1 is the original light intensity of a beam and Φ is the intensity after traversing a thickness x of the solid, then

$$\Phi = \Phi_1 e^{(-\alpha x)}$$

6-30

where α is the absorption coefficient. It may be demonstrated that the absorption coefficient is inversely related to the mean free path. Since $e^{(-\alpha l)}$ is the fraction transmitted through the junction, then $(1- e^{(-\alpha l)})$ is the fraction of the original beam that has been absorbed at distance l.

Not all the minority carriers (minority carriers are holes in n-type material, or electrons in p type material) contribute to power, some of them recombine with majority carriers in the bulk or at the surface. That is why η_{co} is defined as the ratio of the carriers passing thru the circuit (the short circuit current density, J_s) to the total numbers of carriers generated in the solid per unit time. Rearranging 6-29 serves as the definition of collection efficiency:

$$\eta_{co} = \frac{J_s}{(1-r)[1-e^{(-\alpha l)}]en_{ph}}$$

6-31

This equation has been computed for the case of an infinite plane p-n junction at x=l (little el); it is a function of the absorption constant for the radiation , the minority carrier lifetime τ^* , and the surface recombination velocity s. The latter is the effective velocity at which all minority carriers appear to be swept into the surface where they disappear in surface trap energy levels. Surface recombination is evaluated by measurement of the total recombination rate as a function of surface-to-volume ratio. It is very sensitive to surface treatment; high s in ground or sandblasted surfaces, low in polished by chemical etching. Table 6.2 has collection efficiency as a function of recombination velocity s, diffusion length L, and absorption coefficient α, and with l (little el)=10^{-3} cm .

Table 6.2 collection efficiencies.

s (cm/sec)	L (cm)	α (cm $^{-1}$)	η_{co}
0	10^{-2}	10^3	0.61
0	10^{-3}	10^3	0.47
0	10^{-3}	10^6	0.65
100	10^{-3}	10^3	0.61
infinity	10^{-2}	10^3	0.25
infinity	10^{-3}	10^3	0.23
infinity	10^{-3}	10^6	0.001
infinity	10^{-6}	10^3	6×10^{-5}

Looking at that table you find that high collection efficiency occurs when l /L<<1 (little el/L << one) and for which s is low. Thus we chemically etch photocells. For efficient absorption the thickness of the top surface of the cell should be approximately equal to the reciprocal of the absorption constant, l (little el) $\gg 1/\alpha = \alpha^{-1}$, therefore $10^{-6} < l$ (el) $< 10^{-4}$ cm. The minority carrier lifetime can be found if we assume l (el) is 0.1 the diffusion length, L = $(D\tau^*)^{1/2}$. Then if D = 10 cm^2 per sec, the lifetime lies between 10^{-5} sec $> \tau^* > 10^{-13}$ sec (for the minority carriers). It can be shown that s will not reduce the collection efficiency if s/ D<<1/L and s / D << α. With l (little el) =0.1L=α^{-1} for α in the range of Table 6.2 then both of these conditions are true if s is less that 10^5 cm per sec. These reasonable values for τ^* and s mean that there are many usable materials, so collection efficiency does not affect the choice of semiconductors used.

The amount of reflection loss can be approximately determined by the relationship between the index of refraction, γ and the reflection coefficient, r

$$r = (\gamma - 1)^2 / (\gamma + 1)^2$$

6-32

There is an empirical relation that has been found to be true for diamond lattice structure material, such as the semiconductors we are interested in.

$$\varepsilon_g (\gamma)^4 = 173 eV$$

6-33

101

If the energy gap, $\varepsilon_g = 1eV$ then $\gamma = 3.6$ and $r=0.31$ For $\varepsilon_g = 2eV$ then $\gamma = 3.05$ and $r=0.26$. The conclusion is that for materials used for photovoltaics they will all have about the same reflection loss. This simplifies our choices of materials.

We can further simplify eq. 6-23 for efficiency, which will give us an upper bound for the efficiency that ANY cell can operate. We write the short circuit current as $J_s = K n_{ph}$, where K is a constant that contains the losses above (reflection, transmission, and collection). If we assume that K is approximately 1 and that eV_{mp}/kT is much greater than 1 then eq. 6-23 is

$$\eta_{max} \approx \frac{n_{ph}V_{mp}}{N_{ph}\varepsilon_{av}}$$

6-34

For a silicon cell, n_{ph} is approximately (2/3) N_{ph} and the V_{mp} is about 1/3rd the energy of the average impinging photons, e_{av}. The results of these approximations give us the maximum efficiency of a silicon cell of about 22 percent. For all the other semiconductors you will get a little more or less efficiency. However, no choice of materials will give high efficiencies of 50, 60, or 70 percent.

The last microscopic parameter we look at will be the saturation current density J_0. We assume that the junction region is thin in comparison with the diffusion length and so recombination within the junction proper may be neglected. We consider a heavily doped p region and a moderately doped n region, which is favorable for a low value of J_0. Thus, we assume that the current is carried solely by the holes generated in the n region that diffuse over to the p region. The following mouthful will be typed in one continuous sentence, and then put into equation form. The current carried by these holes is equal approximately to the absolute value of the electronic charge, $|e|$, times the equilibrium hole density, p_n, in the n region, times the mean diffusion velocity u_p, which we assume to be the diffusion length $L = (D\tau^*)^{1/2}$ divided by the hole lifetime τ_p^*, or...

$$u_p = \frac{(D_p\tau_p^*)^{1/2}}{\tau_p^*} = \frac{D_p}{L_p}$$

6-35

In germanium at room temp, values of 200 cm/sec are typical for holes having a lifetime of 10^{-3} sec. In silicon diffusion velocities of 10^4 and lifetimes of 10^{-7} are typical. The hole generation current density is the approximately

$$J_{0(p)} = \frac{p_n e D_p}{L_p} = \frac{p_n \mu_p kT}{L_p}$$

6-36

Where the Einstein relation $D = \frac{\mu kT}{e}$ was used. From Eq. 3-44 the product of the hole and electron densities in a given region must be constant

$$p_n = \frac{n_i^2}{n_n} = \frac{2.23x10^{31}T^3 e^{[-\varepsilon_g/(kT)]}}{n_n}$$

6-37

Where n_i = equilibrium density of electrons in the intrinsic semiconductor and n_n is the equilibrium density of electrons in the n region. The latter quantity may be expressed in terms of the electrical resistivity as

$$n_n = (\rho_n e \mu_n)^{-1}$$

6-38

Combining these last 3 equations yields

$$J_{0(p)} = 2.23x10^{31}T^3 \rho_n \mu_n \mu_p kTeL_p^{-1} e^{[-\varepsilon_g/(kT)]}$$

6-39

This shows the strong temperature dependance of J_0. The lower J_0 is, the higher the efficiency, therefore we wish to keep cells at a lower temperature to improve efficiency.

Choice of Materials

Putting this all together, two effects take place. First, the number of photons absorbed with energy greater than the band gap, n_{ph}, decreases as the band gap increases. As the band gap increases, the saturation current density J_0 decreases, thus causing an increase in the output voltage. There is a parabolic effect on efficiency. The efficiency is a maximum at an energy gap of about 1.5 eV.

Cells in space

In space we can't fix the cells, so we design them with a series parallel design to minimize problems. If one cell opens up electrically speaking, you're not dead. The ISS arrays have a diode every 8 cells. If a cell opens, the current is moved through the diode, and we don't lose the entire number of cells in series, only 8. We also have to worry about radiation damage from the Van Allen belts. One way to minimize damage is to maximize carrier mobility. Since electron mobility is two to three times hole mobility, the use of very thin n-type on top of p-type has notably improved the radiation resistance of solar cells. Also, sapphire covers over the cells protects the cells from damage. In one experiment 0.25 gm/cm squared covers stopped 2 MeV electrons. Stopping protons is a lot harder. Gallium-Arsenide (GaAs) solar cells are more efficient at any given temperature than Si and somewhat more resistant to radiation damage--but they are more expensive and massive. Remember lower mass is desirable in spacecraft, so there is a tradeoff between efficiency and mass.

Getting more efficient

There are some basic reasons why solar cells can't reach their theoretical efficiencies.

1. Reflection losses on the surface. This has been largely removed by using antireflective coatings, but still amounts to 3-6%.

2. Incomplete absorption. We want to use materials such as Gallium-Arsenide (GaAs) that have absorption coefficients that increase rapidly with photon energy.

3. Utilization of only part of the photon energy for creation of hole-electron pairs. Photons with too little energy don't create pairs, those with too high energy give electrons KE which is dissipated as heat. For a energy gap of 0.9eV 46% of the impinging solar energy can be utilized for hole electron pairs, assuming all the photons are absorbed. This limit and the number of photons per square meter per second are independent of fabrication technique.

4. Incomplete collection of hole-electron pairs. Pairs generated outside the junction drift together not producing current. There is a technique called back surface field technology where an electric field is created to push randomly moving free carriers back toward the pn junction. Also we can use a metal back to act as a mirror to reflect light back towards the p-n junction. (BSR-Back Surface Reflector) This can give 2-3% increase in efficiency. However, if we grid the back, the cell will operate 10-15 degrees cooler (infrared goes through, IR can't be used by the cell). Lower temperature means increased power rate of about 1.5%/degree. This allows reduction in area of about 8%.

5. Others, such as open circuit voltage.

The previous analysis has only concerned crystalline silicon or other semiconductors. There is something called amorphous silicon (a-Si) which does not have a regular lattice. The amorphous structure increases considerably the probability that light will be absorbed producing a hole-electron pair. In crystalline silicon (called a indirect-band-gap material) two coincident phenomena are required to promote a valence electron to the conduction band: the light absorption itself, and the favorable interaction of a lattice vibration (called a phonon). Amorphous Silicon is a direct-band-gap material: the light energy alone is sufficient. This difference between a single and two event accounts for the order of magnitude difference in the absorption of light by the materials. Therefore a-Si can be manufactured in thin film form for semiconductors (0.5 micrometers vs. up to 300 micrometers for crystalline Si). They will use less material so they will weigh less and cost less.

Another development is amorphous silicon carbide (a-SiC). By varying the amount of carbon you can change the energy gap.

Mechanical ways to get more efficient

Concentrators can be used to put more solar energy per square centimeter, thus getting more power output, a cheap way to increase efficiency. The only concentrator system flown to date was the Hughes Space Communications Company's HS 702 geosynchronous spacecraft. It used two reflectors at about a 45 degree angle to the solar cells to reflect light onto the cells thus concentrating the solar flux. The first one, the Galaxy XI satellite, was launched on 21 December 1999. The solar arrays were deployed on 21 January 2000 and, for the first time, television cameras were used to watch the deployment. Most concentrator designs use a lens to focus more sunlight onto the cells. The Hughes design used reflector panels on each side of the solar panel. The cells were Ga-As and produce 12.4 kw at approximately 25% efficiency. See figure 6.4

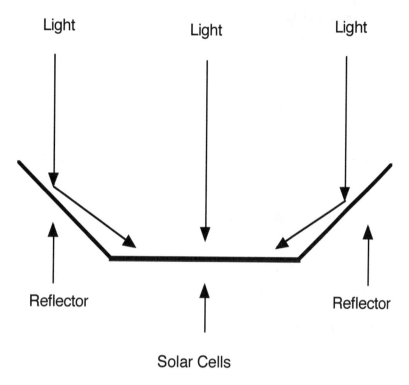

Figure 6.4 Reflector concentration of light

A different concentrator used a fresnel lens to concentrate the solar rays onto the cells. This was the SCARLET project (Solar Concentrator Array with Refractive Linear Element Technology). It was successfully deployed on the Deep Space 1 spacecraft.

Concentrators can be combined with multi-junction cells. Multi-junction cells, also called triple junction cells, stack cells on top of each other, each efficient at absorbing certain band-gap frequencies (colors) of light but allowing other frequencies to be transmitted to the next cell. The top cell might collect blue light, allowing red light to pass on to a cell that is more efficient in that part of the spectrum. A GaAs three-junction multi-band-gap cell has a calculated efficiency of over 34%.

As of October 2016 Spectrolab has produced over 4 million multi junction cells made of Ga,InP/GaAs/Ge semiconductors with efficiencies of 26-30%. Figure 6.5.

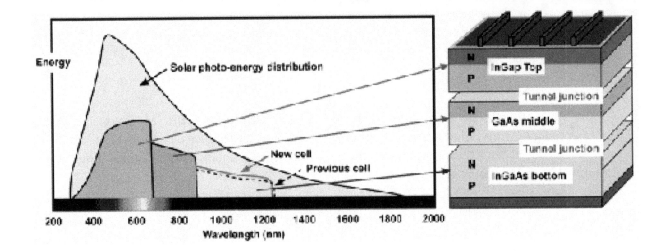

Figure 6.5 Triple junction cells

Another method was proposed in January 1999 by Wayne Phillips (NASA Tech Brief Vol. 23 No. 1 "Solar-Cell system with High Efficiency"). Prisms are placed in front of the light and different Solar Cells are placed under the rainbow. Each cell is chosen to be efficient at that color (frequency) of light. The estimates of efficiency at 1 Sun power is 32% and with concentrated light at 8 Suns of 43%. The prisms can be inexpensive plastic or glass. This has never flown.

Solar Cells are also being developed that convert infrared (heat) to electricity. These are called ThermoPhotoVoltaics (TPV). This could allow the radiators on the Space Station to have TPV Cells to produce just a little more power. Terrestrial applications are even more practical--remember geothermal power?

Efficiency of Photovoltaics

In general, the efficiencies are about 15 to 20% with multi junction cells at 30% or more.

Problems

1. For a material with a surface recombination velocity of 0 cm per sec, a diffusion length of 10^{-3} cm, an absorption coefficient of 10^3 cm^{-1}, a depth of surface of n-type material of 10^{-3} cm, an energy gap of 1.5 eV, and which is irradiated by light (possessing energy in excess of the energy gap) of 5×10^{17} photons per second per cm^2, find the short circuit current density.

2. A certain thin film CdS cell had the following properties at 55° C: short circuit current density=15 ma/cm^2; max power current density=11.5 ma/cm^2; open circuit voltage=0.45 V; voltage at maximum power= 0.34 V; efficiency = 3.3%. The phase of the moon is one quarter. Joe is smoking a pipe. In other words, not all this data is necessary. A space probe needs 800 Watts at 28 Volts. Each cell is 2 cm by 6 cm.

a) Design a power system from these cells which will supply the required power. Think of these cells as a bunch of batteries that you have to hook up in series and parallel. Draw the system, labeling everything. This is a system, not a cell design.

b) What is the minimum area required?

3. The area of the solar array on the ACTS spacecraft is 144 sq. ft. The power output is 1400 Watts. Assume an incident radiation is 0.135 W/cm^2

a) Calculate the efficiency of the array.

b) If the incident radiation was 0.85 W/cm^2, what would be the output power?

Chapter 7, Thermionic and Magneto-Hydrodynamic Generators

Thermionic Generators

In simple language, here's how a thermionic converter works. The emitter is at T_e (T hot) and the collector is at T_c (T cold). The emitter has a fermi level that is lower than the collector's. The work function, ϕ_e that is required to lift the emitter's electron from the fermi level to the infinite reaches of outer space (i.e. so that the electron leaves the atom) is greater than the collector's work function, ϕ_c.

When the emitter is heated, some of the electrons gain enough energy to leave the emitter and they travel to the collector. There they give up their kinetic energy plus the collector's work function in order to enter the collector. This extra energy is taken away as heat. The electron still has extra energy because $\varepsilon_{f_c} < \varepsilon_{fe}$ (Ve>Vc), and this can be used to drive a load.

A complete circuit is shown in figure 7.1.

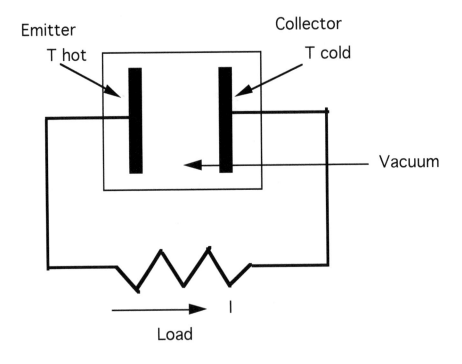

Figure 7.1 Thermionic Generator

Once the emissions of electrons become significant, a space charge layer builds up between the emitter and collector. This is a cloud of electrons in front of the emitter. The magnitude of the space charge layer depends upon the emission current and $\Psi_c - \Psi_e$. Obviously, the space charge layer reduces efficiency because the electrons must overcome this potential difference. The key to getting efficient converters is to reduce the space charge layer. There are many ways of doing this, but the preferred way is adding Cesium. The Cesium ionizes, creating a positively charged atom that attracts the electrons, much like the grid on a vacuum tube. Yes, vacuum tubes still exist.

For space power systems, the most promising ways of using thermionics would be either an in core or out of core reactor system. The in core has the thermionic generators in the core of the reactor, while the out of core has a heat transfer system, such as a pumped fluid loop or heat pipes, to take the heat to the thermionic generators. The former has an estimated efficiency of 11%, while the latter would have 15%. The latter is also thought to be easier to build.

The Soviet Union flew two TOPAZ 5 Kw electric nuclear reactors with in core thermionic generators. They could generate that power for 3-5 years from 12 kg of fuel. The reactor mass was 320 kg. These had operating temperatures of T_H at 1800 to 2000 K and T_C at 800 to 1000 K. They flew in 1987 and 1988.

Efficiency of Thermionics

In general, the efficiencies are about 10 to 15%
Add PETE pete solar technology search on google.

Magneto-Hydrodynamic Generators

Simply put, MHD's are generators that use an ionized gas flowing through a magnetic field to generate electricity. So, first we must talk about how to ionize gases. There are three basic ways: Thermal, i.e. collisions, irradiation, i.e. photon collisions, and cumulative, i.e. ionization by stages. For MHDs thermal is the most useful.

Thermal Ionization

Consider a molecule of oxygen. It can be represented by a dumbbell. When you heat it to 3000 degrees K the bonds that hold the two atoms break. If you continue to heat it the electrons get enough energy to break free from the atom and we get ions. For oxygen at atmospheric pressure, this happens at 10,000 K. This ionization is dependent on pressure, at 1×10^{-6} atmospheres it occurs at 6000 K. This is one form of thermal ionization.

When molecules with high energy bump into each other enough energy can be transferred between the neutral molecules to ionize one of them. This is another type of thermal ionization. This can occur only at very high temperatures.

Conductivity

Once we have a gas ionized it can conduct current. We want a high conductivity to conduct current really well. Using equations 7-1 and 7-2 one can calculate that for argon at 7000 K and 1 atmosphere the thermal conductivity is 6.72 per ohm-cm (a resistivity of 0.1488 ohm-cm), which is pretty good. Unfortunately present materials can only take about 3000 K. So, we must find another way to get good conductivity. One way to increase conductivity is by seeding, i.e. adding a substance with a low ionization energy such as alkali compounds. The fraction of ionized atoms is X, then

$$\frac{X^2}{1-X^2}P = 3.16x10^{-7}T^{\frac{5}{2}}e^{[\frac{e_i}{kT}]}$$

7-1

$$\sigma = 3.34x10^{-12}\frac{X}{QT^{\frac{1}{2}}}(ohm-cm)^{-1}$$

7-2

Q is the collision cross section, P is the total pressure in atmospheres, T is the temperature in Kelvin, and e_i is the ionization energy.

A quick and dirty MHD analysis

Figure 7.2 shows a typical MHD generator.

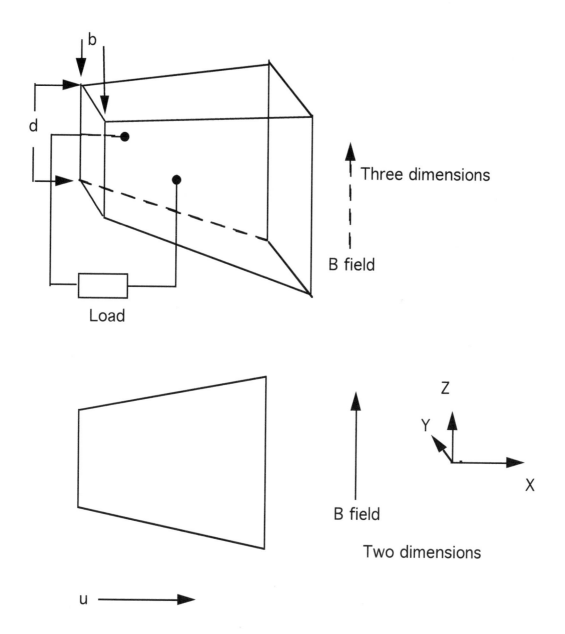

Figure 7.2 MHD generator

The gas moves at constant velocity u in the x direction with a magnetic field B at right angles. The Hall effect creates an electric field given by

$$\vec{E}_{ind} = \vec{u} \, X \, \vec{B} \quad ind \; is \; induced \tag{7-3}$$

The gas has a conductivity, therefore a current density is induced given by

$$\vec{J}_{ind} = \sigma \vec{E}_{ind} \tag{7-4}$$

Simultaneously the induced current interacts with the magnetic field to induce a Lorentz force

$$\vec{F}_{ind} = \vec{J}_{ind} \, X \, \vec{B}$$

7-5

This acts in the opposite direction of $\vec{u}$, thus slowing down the gas. So kinetic energy is transformed into electrical energy. But, we have an expanding nozzle which speeds up the gas, maintaining a constant speed. As the gas expands, it cools. Therefore, the net result is that the temperature of the gas is converted into electrical energy directly. This is a simple explanation, in reality it is more complex.

The generator we have described works fine, but the Hall effect produces axial current in the plates, which reduces the power density. If we use many electrode segments, such as in figure 7.3, we eliminate the axial current, and the power density goes up.

Segmented Generator

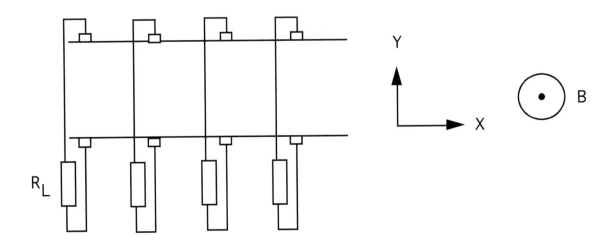

Figure 7.3 Segmented Generator

The efficiency is given by dividing the electrical power output by the mechanical power input, which in these two cases is the same. This is sometimes called a loading factor, K.

Segmented generator efficiency

$$K_H = -\frac{E_x}{\beta_n uB}$$

7-6

Where β_n and K_H are the Hall parameter and the Hall loading factor, both dimensionless.

A third way to make an MHD is shown in figure 7.4.

Hall Generator

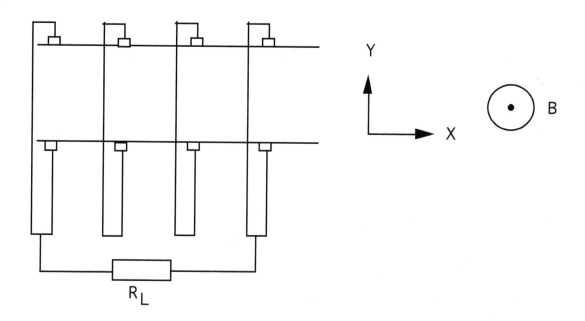

Figure 7.4 Hall Generator

The middle segments are shorted together. This short circuits the transverse current and power is collected by the upstream and downstream electrodes. This is called a Hall generator.

Hall generator efficiency

$$\eta_e = \frac{\beta_n^2 K_H (1 - K_H)}{1 + \beta_n^2 K_H}$$

7-7

$$\beta_n = \mu_n B = \omega\tau$$

$$\omega = eB / m^*$$

$$\mu = e\tau / m^*$$

7-8

Where m^* is the effective mass of the electron, τ is the mean time between collisions (seconds), e is the electric charge (coulombs), ω is the electron cyclotron frequency (1/seconds).

$$K_H = -\frac{E_x}{\beta_n uB}$$

7-9

One can maximize K_H and get a theoretical maximum efficiency, which is given by equation 7-10

$$\eta_{e(\max)} = 1 + \frac{2}{\beta_n^2} - \frac{2}{\beta_n}(1 + \frac{1}{\beta_n^2})^{\frac{1}{2}}$$

7-10

The results of all this is that for $\beta_n >$ about 0.7, the segmented generator is the most advantageous. We generally ignore the mobility of ions because $\mu_n >> \mu_{ions}$.

Also, the temperature of the exhaust gases are on the order of 2500 K, thus you can use the gases coming out the exit as a source of energy for a conventional power plant. (a Navy boiler might have steam at 686 F [636 K] for instance). The efficiency of an MHD generator is approximately 55%.

Problems with MHD power generation

Seeding material, such as alkali metals, can corrode the components of the generator. There is a trade off of amount (cost) of seed material vs. amount of material you can re-use.

It is hard to manufacture high temperature materials that are conductors or insulators. Things tend to melt at high temperatures.

Magnetic field intensities should be high for greater power. This leads to using superconducting magnets which means we need low temperatures. For space applications supplying cryogenic liquids to LEO could be very costly. Higher temperature superconductors could help this.

The MHD principle is fully reversible. Not only was it used in the novel, Hunt for Red October, but the Japanese made a ship that moved in a bay (as in a body of water) using MHD in the 1990's. This ship had cryogenically cooled magnets and had a speed of about 8 knots. However, no MHD power supply has flown in space.

Efficiencies of MHDs

Although they have never flown in space, they are described here as an example of what could be developed. Their efficiencies could be 55%.

Problems

1. A Hall generator (MHD) has the following characteristics.
E_x = -1200 V/m
B = 3 tesla
u = 790 m/s
β_n = 1.25
d = 2 m
b = 0.6 m

Find the efficiency of the generator, be careful.

2.

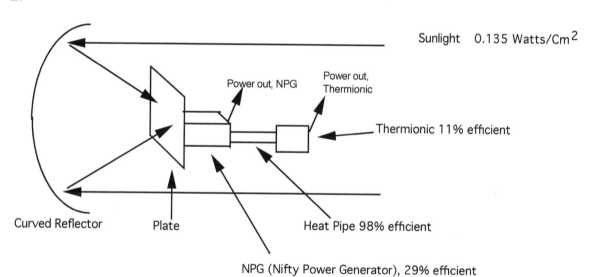

The plate has an area of 10 m². Curved reflector collects 100 m² of light and concentrates it to plate (10 to 1 ratio).

a) How much electrical power does the NPG put out?
b) How much electrical power does the Thermionic put out?

c) What is the efficiency of the <u>System</u>? $\eta = \dfrac{Pout(total)}{Pin}$

115

Chapter 8, Carnot, Rankine, Brayton, and Stirling Cycle Generators

For more powerful Space Power Systems, we can learn from Terrestrial systems, namely, Brayton, Rankine, and Stirling. These all are limited in efficiency by the Carnot cycle,

$$\eta_C = 1 - \frac{T_C}{T_H}$$

Carnot Cycle

Figure 8.1 shows the ideal or conceptual Carnot cycle engine proposed by Nicolas Leonard Sadi Carnot in 1824.

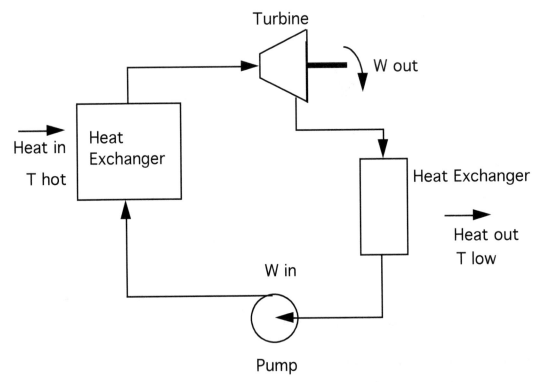

Figure 8.1 Carnot cycle (conceptual)

Heat is added to the Heat Exchanger which heats the working fluid and increases its pressure. The hot fluid moves into the turbine, rotating it, which generates the Work (out) and decreases the pressure. The fluid goes to the second Heat Exchanger and is cooled. It then goes to a pump, which increases the fluids pressure and send it back to the heat exchanger.

Rankine Cycle

Figure 8.2 shows the Rankine cycle, named after a Scotsman, William John Macquorn Rankine.

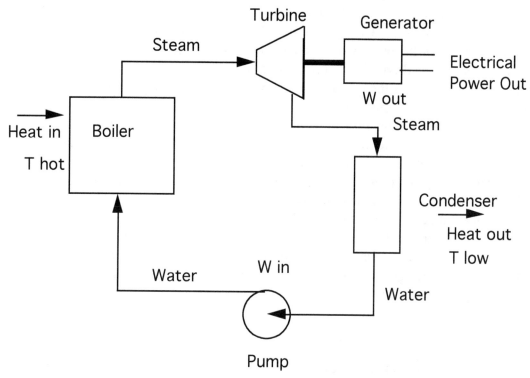

Figure 8.2 Rankine Cycle (ideal)

This is the most common method on Earth to produce power from heat. It is used by power companies and ships at sea. Water is heated in a boiler and turned into steam. The steam is very dry, that is, it is molecular water. It is not clumps of many water molecules, it is H2O in single molecules. If the steam were in droplet form when it was put into the turbines, the energy contained in the drops would tear the turbine blades off the turbine shaft. This would cause the destruction of the turbine. The steam hits the turbine blades and turn the turbine. The turbines are connected to a generator, which turns producing electricity. The steam is then condensed back into water in the Condenser, and then sent back into the boiler via the water pump. How does the condenser condense the steam back into water? Usually in terrestrial applications a river or a stream supplies cooling water. On a ship, the ocean supplies cooling water. In space the condenser would be a radiator as well, radiating the heat into space and making the steam condense into water. There are ways to make this cycle more efficient using reheat, regeneration, and superheat, among others, but we won't get into these.

In terrestrial plants using water, the highest temperature is dependent upon the material (typically steel) and is roughly 570 C. The cold temperature is typically 30 C. Using these temperatures the Carnot efficiency would be 63%, but the actual efficiency for a Brayton cycle would be 42%. But that doesn't consider the pump power needed to pump the water, the transformer efficiency, the mechanical losses of the turbines, and other things. So the final efficiency is about 30 to 33% in converting heat to electricity.

As an example of the temperatures and pressures involved, the author was on an aircraft carrier that had boilers. The steam leaving the boilers was at 363 C, 686 F at 41.37 Bar, 600 Psi. That means that the water pump returning the water to the boiler must pressurize to greater than 600 Psi or the water won't go back into the boiler. In order to reach these pressures, the pump may be driven by steam (more efficient), or may be electric (emergency backup). All of this lowers the overall efficiency. There are some problems in practical systems that complicate the engineering design. No system is completely closed, so water must be replaced. A problem is that air can be introduced into the system. Both of these are not good for a Space System, and must be greatly reduced in the design, thus driving up cost of design and testing. The oxygen component of air at these temperatures and pressures can severely corrode the pipes in the boiler. So de-aerating systems are used to remove the air from the water before it enters the boiler (not pictured). Finally the water level in the boiler must be maintained at just the right level. Too low a level and the pipes in the boiler are not cooled by the water and melt, causing an explosion. Too high a level and moisture is carried over into the turbine causing turbine blades to fly everywhere. Both can ruin the system and make a space power design difficult. This is why Rankine systems with water are not designed for space power systems.

Brayton Cycle

Figure 8.3 shows the Brayton cycle, named after and developed, but not invented by an American, George Brayton.

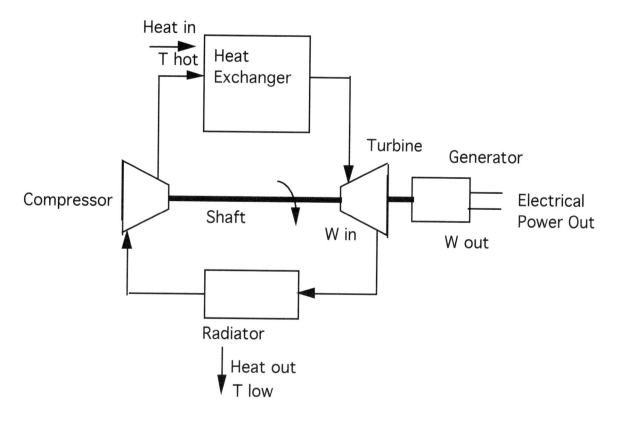

Figure 8.3 Brayton Cycle (ideal)

This is similar to Rankine except a gas is heated, sent through a turbine which turns a generator as is done in a Rankine Cycle. The gas then goes to a radiator to cool, then to a compressor (powered by a shaft hooked to the turbine) which builds up enough pressure to send it back into the Heat exchanger. Again, to make it more efficient you can have regeneration, reheat, inter-cooling or other methods. The advantage in using this in Space Power systems is that there is no phase change. The problems with having a liquid impinge upon turbines and tearing the turbine blades off the turbine shaft are eliminated. Also, water has a property that when it freezes it expands. Expanding water breaks pipes on Earth, and breaking pipes in outer space would be disastrous. (On the Space Station, electric heaters are used to ensure that the water side of the cooling loop never freezes). Using a gas eliminates this problem as well. Typical gases that are used are inert, such as He or Ne or mixtures of the two to avoid corrosion of the pipes and other components.

Stirling Cycle

The Stirling cycle was invented by Reverend Dr. Robert Stirling in 1816, another Scotsman. It is really weird, it uses pistons to move up and down. No rotation, like the above systems, more like the up and down motion of an automobile engine being converted to rotation. However, there is much research on using free piston Stirling engines for space power. They usually use two pistons on opposite sides to reduce vibration. Here is a slide from the Glenn Research Center at NASA that shows the cycle of the free piston Stirling. These generators were planned to be used in space.

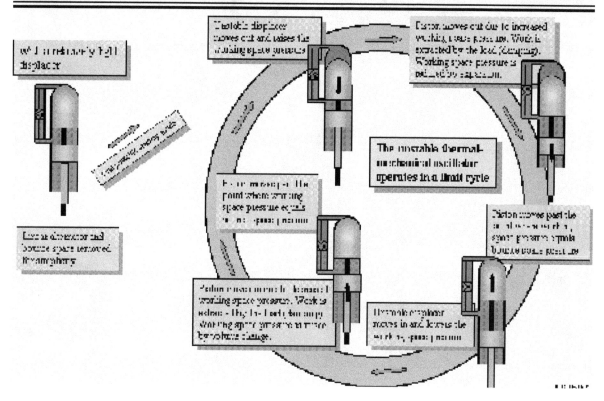

Figure 8.4 Stirling Generator (Courtesy NASA) http://www.grc.nasa.gov/WWW/tmsb/stirling/intro_stirling/

The piston moving back and forth would be connected to a generator to produce power. To eliminate vibration, most designs have two generators connected together as shown below. The advantage over Brayton and Rankine is that there are no turbines, compressors, or pumps. The disadvantage is that it still has moving parts. Lubrication in all three systems is an engineering problem to be solved.

Figure 8.5 Dual Stirling Generator (Courtesy NASA)

Efficiencies

As we said in the first chapter, the efficiencies of converting heat into electrical power for these systems are roughly 30%. This is a sizable efficiency and is why for the Space Station there have been proposals for a Solar Brayton Cycle generator. More on that in the Space Station chapter.

DIPS (Dynamic Isotope Power Systems)

Dynamic Isotope Power Systems, which is basically an easy concept. For power of up to about 1 Kwe, you can use a radioisotope to power a Rankine, Brayton, or Stirling device. This obviously has the advantage of 3 times the efficiency (roughly) of a TEG. NASA was testing Stirling power systems for use in DIPS at the Glenn Research Center. The engineering Unit was assembled in 2008. The program was cancelled in November 2013 due to budget constraints.

http://www.planetary.org/blogs/casey-dreier/2013/20131115-nasa-just-cancelled-its-asrg-program.html

Temperature Control

Temperature control is a textbook all by itself. Temperature control on any Spacecraft, be that a satellite, rover on a planet, or the Space Station, is a very complex topic. The biggest thing to note is that Space is neither hot nor cold. In Earth orbit, the spacecraft is really hot when exposed to the Sun, and really cold when in the Earth's shadow. Every electrical device on the spacecraft produces heat. Power in equals power out plus heat out. When designing Space Power Systems, heat producing devices (electronics, power converters, power supplies, pumps, etc.) must be cooled. Also, keeping the Sun's heat out is important. If you send a spacecraft to Mercury, your big problem is keeping the Sun's heat out, while cooling your heat generators. If you send a craft to Pluto, your problem is to keep the heat in to keep electronics, valves, fuel, etc. within normal operating limits. That's why the Galileo probe and the Mars Spirit and Opportunity rovers had Radioisotope Heating Units (RHU's). RHUs are small devices (2.54 cm by 3.3 cm--1 by 1.3 inches) that supply one Watt of heat (thermal). They use the heat of radioactive decay of Plutonium Dioxide to provide the heat and only weigh about 40 grams (1.4 ounces). The Plutonium is about the size of a pencil eraser.

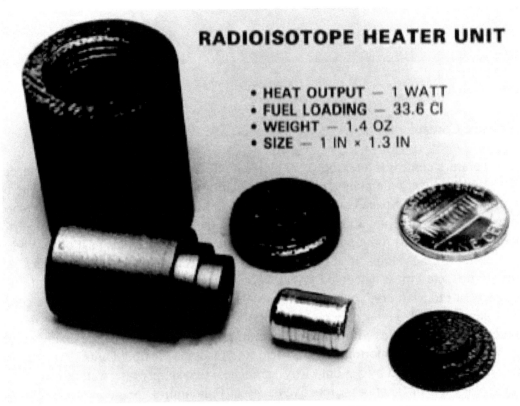

Figure 8.6 RHU (Courtesy the Department of Energy (DOE))

LIGHTWEIGHT RADIOISOTOPE
HEATER UNIT

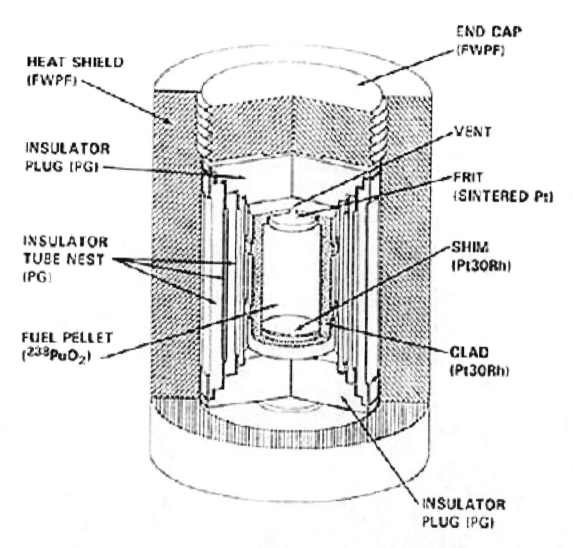

Pt = Platinum
Rh = Rhodium
PG = Pyrolytic Graphite

$^{238}PuO_2$ = Plutonium Dioxide
FWPF = carbon-carbon composite
woven with perpendicularly oriented
graphite fiber

Figure 8.7 RHU (Courtesy NASA)

Heating and cooling can be accomplished with active or passive systems. Active are things like the ammonia system on the International Space Station, louvers and electrical heaters with thermostats, evaporative coolers (such as flash evaporators on the Shuttle). Louvers has a bimetallic coil, (like on older air conditioning thermostats) which operates a spring, that opens the rectangular blades, like a venetian blind. It requires no power.

Passive are things like Multi Layer Insulation (MLI), aerogel, putting heat producing devices next to devices that need heat, reflective or absorptive paints, RHU's, and heat pipes.

Figure 8.8 MLI. The white are separator layers to keep heat conduction between layers small

Heat Pipes

Heat pipes can transfer a great deal of thermal energy from one point to another.

Figure 8.9 shows a heat pipe.

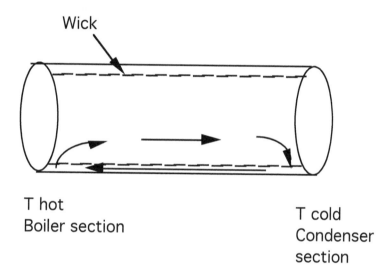

Figure 8.9 Heat Pipe

Here's how it works, Heat enters the boiler section, the working fluid evaporates and travels to the condenser, where it gives up its latent heat. In the wick, surface tension forces return the condensate back to the evaporator section through capillary channels. Using the latent heat of evaporation/condensation is extremely efficient--90% or so. Heat pipes can transfer up to 500 times as much heat per unit weight as can a solid thermal conductor of the same cross section. In theory, the heat pipe operates almost at a single temperature, in reality a small temperature gradient exists, because a small vapor pressure gradient is generated between the boiler and condenser sections. They can be designed to operate with a temperature difference of only a few degrees. In practice, 5 kW thermal have been driven down a 2 ft. molybdenum heat pipe at 1700 K with a temperature drop of 6 K. The heat pipes don't have to be cylindrical, they can be rectangular, square, they can start on one end as square, and the other end as rectangular. Whatever is needed to fit the device to be cooled. They do not have to be straight, they can curve around corners. You will see heat pipes all over the place, several experiments have flown on the Shuttle, a candy company uses them to make candy canes more efficiently, and home air conditioners are being built with heat pipes to be more efficient. On the Space Station, the computers for Node 1 are external and are on the Primary Mating Adapter #1. Both computers have Heat Pipe Radiators to cool them. On the P6 segment of the Space Station (a solar array section), the DC to DC Conversion Unit

(DDCU) is mounted on a heat pipe radiator as well, which are exterior to the station, exposed to space. Multiple pipes are used for meteor protection--if one pipe is hit and all the fluid leaks out, other pipes can still dissipate heat.

Chapter 9, Introduction to Radiation

The first thing to talk about is radioactivity. It is defined as the process of spontaneous decay of unstable atoms. These unstable atoms emit one or more of the following nuclear radiations.

Alpha Particles are essentially helium atoms minus the electrons. Two protons and two neutrons doubly charged. They are the least penetrating of the radiations and can be stopped by a thin sheet of paper. Their penetration through air is roughly 9 cm (3 inches or 9 barleycorns). If alpha emitters are ingested, they are very dangerous to a persons health (more on this later).

Beta particles are electrons or anti-electrons. The electron is negatively charged, the anti-electron positively charged. They are slightly more energetic, being stopped by thin metal or clothing. They can penetrate 10 meters (30 feet) in air. Beta radiation can cause skin burns and beta emitters are also dangerous if ingested.

Gamma Rays are high frequency electromagnetic radiation. They are extremely energetic, requiring thick tungsten or lead to attenuate them. They frequently accompany alpha and beta emissions and always accompany fission. (Figure 3.3)

Neutrons are uncharged elementary particles normally found in the nucleus. A neutron has a mass slightly larger than the proton. A free neutron is unstable and changes into a proton, a beta particle, and an electron antineutrino $\bar{\nu}_e$. This reaction has a half life of 12 minutes. For our studies of reactors this is of no consequence, since the lifetime of a neutron in a reactor is 0.001 seconds. Neutrons make nuclear reactors possible because they keep the reaction going. They are very penetrating because they are uncharged and require special shielding.

Neutrinos are right up there with invisible massless elephants. It is a zero rest mass electrically neutral particle which comes in six flavors, only two of which, the electron neutrino and electron antineutrino, are of concern to nuclear reactors. They are very penetrating, solar neutrinos pass right through the entire planet without interacting with much matter.

Protons are what give an atom its identity. It is a positively charged particle in the nucleus, and the number of protons equals the atomic number, Z.

Nuclide Designations

A nuclide is an atomic nucleus characterized by its number of protons, its number of neutrons and its excited state, and other things. We will use the following notation for nuclides:

$$^{400}_{180}Bc^{++}_{220}$$

This is the author's invented element, Beccium. The atomic number is the left subscript, the mass number (the total of the neutrons and protons) is the left superscript, the ionization is the right superscript, and the number of neutrons is the right subscript. The right subscript is often omitted, since you can subtract the other two numbers to get its value.

Atomic Weight

The atomic weight of an atom is the mass of an atom relative to other atoms. The atomic mass unit (amu) is defined as one twelfth the mass of a neutral atom of the most abundant isotope of carbon. 1 amu = 1.66053 10^{-27} kg. An isotope is an element that has the same atomic number (number of protons in the nucleus) but has a different atomic mass. The mass of an atom is approximately equal to the mass of its protons and neutrons (because the mass of the electrons is negligible). Therefore, the mass of an atom is numerically equal to its atomic weight.

Mass and Energy Relationships

Nuclear energy obeys Einstein's famous equation.

$$E = c^2 \ m_0 \hspace{6cm} \text{9-1}$$

c = 2.9979 X 10^8 m/s (the speed of light)
m_0 = rest mass.
If we use the MKS system the units are in Joules. The electron volt is equal to 1.60219 X 10^{-19} Joules. Using 9.1, the atomic mass unit, (1.66053 X 10^{-27} kg) has an energy equivalence of 931.481 MeV. The three basic particles have masses of:

free proton   = 1.00759 amu
free neutron = 1.00898 amu
free electron = 0.000549 amu

Because reality is relative, equation 9.2 comes into play.

$$m = \frac{m_0}{\sqrt{1 - \dfrac{v^2}{c^2}}}$$

$$\text{9-2}$$

where v is the velocity of the particle. The total energy ET is the sum of the rest mass energy and the kinetic energy.

$$E_T = mc^2$$

9-3

The kinetic energy is

$$E_{KE} = E_T - E_0 = mc^2 - m_0c^2$$

9-4

$$E_{KE} = \left[\frac{1 - \sqrt{1 - \frac{v^2}{c^2}}}{\sqrt{1 - \frac{v^2}{c^2}}} \right] m_0c^2$$

Fortunately, if v <= 0.2c this reduces to the well known Newtonian formula

$$E_{KE} = \frac{1}{2}m_0v^2$$

9-5

(although it takes a lot of math to prove this)

Zero Rest Mass Particles

Equations 9.4 and 9.5 only apply to particles that have a mass when at rest. For zero rest mass particles we have the famous equation

$$E = h\upsilon$$

9-6

where h is Planck's constant, 6.626196 10-34 joule-seconds (or 4.136 10-15 eV-sec) and υ is the frequency in Hertz. Zero rest mass particles travel at the speed of light. They are considered to be wavelike in nature. Remember, the above equation can only be used for particles with no mass!

There is another equation called de Broglie's equation which relates the wavelength to the momentum of a particle. (Momentum means it has mass = m.)

$$\lambda = \frac{h}{p}$$

9-7

(same equation as 3-15)

Now, if the particle has nonzero rest mass,

$$p = mv$$

9-8

For non-relativistic particles 9.7 becomes

$$\lambda = \frac{h}{\sqrt{2m_0 E}}$$

9-9

For relativistic particles

$$\lambda = \frac{hc}{\sqrt{E_t^2 - E_0^2}}$$

9-10

If the particle has zero rest mass

$$p = E / c$$

9-11

and 9.7 becomes

$$\lambda = hc / E$$

9-12

(same equation as 3-2)

which, in meters, is

$$\lambda = 1.24x10^{-6} / E(ev)$$

9-13

Radioactive Decay

Unstable nuclides will try to become stable by emitting particles or electromagnetic radiation (EM or EMR). In many cases, the product or daughter nuclide is in an excited state and will, in 10^{-15} seconds or so, emit a gamma ray. This is what radioactive decay is all about.

When an alpha particle is emitted, the atomic number decreases by 2 and the mass by 4 (2 protons and 2 neutrons). A typical example is the decay of Plutonium 238 in figure 9.1

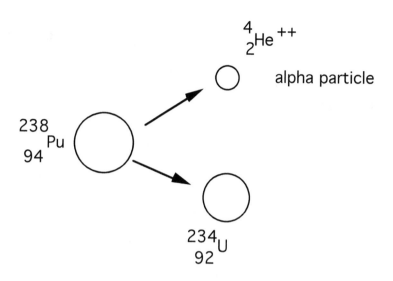

Figure 9.1 Plutonium 238 decay

If the nucleus contains too many neutrons (or two few protons, however you want to look at it), a neutron in the nucleus will decay into a proton and emit an electron and anti-electron neutrino (an antimatter neutrino).

$$_0^1n \rightarrow {_1^1}p + {_{-1}^0}\beta + \bar{v}_e$$

9-14

The bar above v_e means it is antimatter.

The atomic number has increased by one, and its mass has stayed the same. (figure 9.2, which also shows how Plutonium is made) The daughter is more stable than the parent, but if not totally stable, another beta decay can occur. See figure 9.3. T 1/2 is the half life of the atom, which will be defined later.

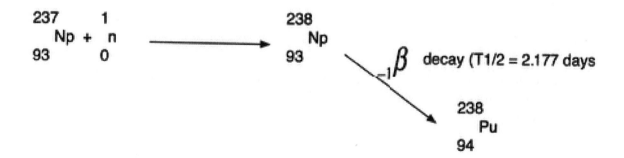

Figure 9.2 Beta minus decay

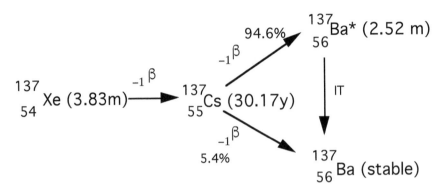

IT is Internal Transition

3.83m is T 1/2 of 3.83 minutes

Figure 9.3 Two Beta minus decays

If the nucleus has too few neutrons, or too many protons, a proton is converted into a neutron and an anti-electron and electron neutrino are emitted from the nucleus.

$$_1^1 p \rightarrow {}_0^1 n + {}_1^0 \beta + \nu_e \qquad\qquad 9\text{-}15$$

The atomic mass remains the same, but the atomic number decreases by one. This cannot happen unless the total energy difference between the parent and daughter is 1.022 MeV. The parent produces one electron rest mass that combines with the proton to make an neutron, and produces another electron rest mass that is the anti-electron. The total energy required to produce both "electron" masses is 1.022 MeV.

The third beta decay process is called electron capture (EC) decay. An orbiting electron is captured by the nucleus, finds and combines with a proton, thus converting the proton to a neutron.

$$_1^1 p + {}_{-1}^0 e \rightarrow {}_0^1 n + \nu_e \qquad\qquad 9\text{-}16$$

This process competes with the previous one when the energy is > 1.022 MeV. Again the atomic mass remains the same, but the atomic number decreases by one. All three beta decays are shown in figure 9.4.

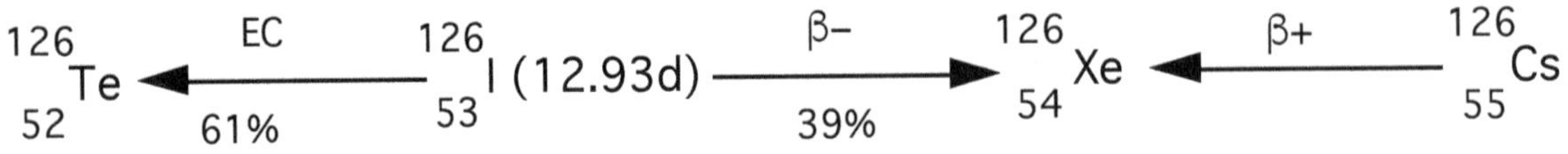

$${}_{52}^{126}\text{Te} \xleftarrow[61\%]{EC} {}_{53}^{126}\text{I (12.93d)} \xrightarrow[39\%]{\beta-} {}_{54}^{126}\text{Xe} \xleftarrow{\beta+} {}_{55}^{126}\text{Cs}$$

Figure 9.4 All three types of Beta decay

In either alpha or beta decay, the daughter nucleus is excited and decays either by gamma emission or internal conversion electron (ce) emission. In the latter, the left over energy is transferred directly to an atomic shell electron, which is ejected from the atom.

In both electron capture and internal conversion (EC and ce) there is a vacancy in an inner shell (inner orbit of the electron). Outer electrons fall into the inner orbit and X-rays or Auger electrons are emitted. The X-ray is emitted at the same energy as the energy difference of the two orbits. An Auger electron is an outer shell electron that is ejected simultaneously with the filling of an inner shell. Thus, the atom is left with two shell vacancies. There can be a cascade effect of X-rays. Eventually, there will only be missing electrons in the outermost shells.

The Law of Radioactive Decay

The law of radioactive decay is as follows:

$$dN / dt = -N\lambda \qquad\qquad 9\text{-}17$$

N is the number of atoms of a particular radio-nuclide
λ is the decay constant (1/time)
t is time

$N\lambda$ is called the activity and is the number of disintegrations per second (dps). There are two measures of radioactivity, the Curie (Ci) and the Becquerel (Bq). The former is approximately the activity of one gram of radium. The latter is the SI system and is defined as one disintegration per second.

$$1 \text{ Ci} = 3.7\text{x}10^{10} \text{ disintegrations/second} \qquad\qquad 9\text{-}18$$

$$1 \text{ Bq} = 2.7027\text{x}10^{-11} \text{ Ci} \qquad\qquad 9\text{-}19$$

You can write 9.17 in integral form as

$$\int \frac{dN}{N} = -\int \lambda dt \qquad\qquad 9\text{-}20$$

If we assume that λ is constant and independent of both the age of the radio-nuclide and the number of nuclides we can integrate this to obtain

$$\ln \frac{N}{N_0} = -\lambda t \qquad\qquad 9\text{-}21$$

This is more commonly put into exponential form as

$$N = N_0 e^{-\lambda t}$$

where N_0 is the number of atoms at $t = 0$.

The half life is just the time it takes for half the original amount of radio-nuclide to decay, i.e. $N = 0.5 N_0$. Half life is the most common way of referring to decay time.

$$T_{\frac{1}{2}} = (\ln 2) / \lambda \text{ or approximately } 0.693 / \lambda$$

Example 9-1:

We have a fuel capsule containing 10 kg of PuO_2 which has a Beginning of Mission (BOM) power of 0.4 watts (thermal)/gram. What is the thermal power output after 10 years? $T_{1/2}$ for Pu is 87.75 years.

$$\lambda = \frac{\ln 2}{T_{1/2}} = \frac{0.69315}{87.75 y} = 7.9x10^{-3} \text{ / year}$$

Heat generated at start of mission $= Q(0) = (10{,}000gm)(0.4W/gm) = 4000$ W thermal. Heat at 10 years is:

$$Q(t) = Q(0)e^{-\lambda t}$$

$$Q(10y) = 4000W_{th} e^{-(7.9x10^{-3} y^{-1})(10y)}$$

$$Q(10y) = 4000(0.924)$$

$$Q(10y) = 3696W_{th}$$

So after 10 years the energy is down by less than 10%. Remember, this is exponential, so after 70 years it won't be close to 70%.

Decay Chains

Daughter nuclides can also decay. Let's look at element W decaying to X (daughter) decaying to Y.

$$W \xrightarrow{\lambda_W} X \xrightarrow{\lambda_X} Y (stable)$$

Where λ_W is the decay constant for the parent nuclide, and λ_X is for the daughter. The general differential equation that describes the growth and decay of the daughter is

$$\frac{dX}{dt} = \lambda_w W - \lambda_x X$$

9-23

Where dX/dt is the time rate of change of the number of daughter nuclides X
$\lambda_w W$ is the number of daughter nuclides being created per unit time
$\lambda_x X$ is the number of daughter nuclides decaying per unit time.

If you assume that there are no daughters at time equal zero, and the initial source of W is W_0 at time equal zero then 9-23 has a solution

$$X(t) = W_0 \frac{\lambda_W}{\lambda_X - \lambda_W} \left[e^{-\lambda_w t} - e^{-\lambda_x t} \right]$$

9-24

If, however, there are daughters at t equal zero; $W=W_0$ and $X=X_0$ at t=0 then the solution to 9.23 is

$$X(t) = X_0 e^{-\lambda_x t} + W_0 \frac{\lambda_W}{\lambda_X - \lambda_W} \left[e^{-\lambda_w t} - e^{-\lambda_x t} \right]$$

9-25

which just adds the decay of X per time (look at equation 9-24).

If the decay chain is short, as in

$$W \xrightarrow{\lambda_w} X(stable)$$

Then X is given by

$$X(t) = W_0 (1 - e^{-\lambda_w t})$$

9-26

Where W_0 is the initial amount of the parent.

Helium Gas Buildup

Anything that emits alpha particles is a problem because the particles will steal two electrons from the surrounding material and become helium. This will build up pressure in an enclosed device, such as a fuel capsule. Equation 9-26 can be rewritten to reflect the helium gas buildup.

$$N_\alpha(t) = N_0 \left[1 - e^{-\lambda t} \right]$$

9-27

Where N_α (t) is the number of helium atoms in the device at time t. N_0 is the initial amount of the alpha emitter.

Radiation Interaction With Matter

The design and operation of reactors, RTGs, and their shields depends on how the radioactive particles (or waves) interact with matter.

Ionization is merely the stripping of an electron (or more than one) from the atom as the radiation goes by the atom. The energy required to do this increases as the atomic number Z increases. Different forms of radiation are better or worse at ionizing. One measure of this is the specific ionization, SI, which is the number of ion pairs created per path length (ip/cm).

There are two types of ionization, direct and indirect. Direct is caused by charged particles and atoms interacting with the orbital electrons. Indirect is caused by neutrons and EMR colliding with atomic nuclei or electrons. (EMR are x and gamma rays). The charged particles liberated in these collisions then travel about and interact with other nuclei or other atom's orbital electrons.

Excitation occurs when the energy is not enough to ionize, but is enough to put the electron in a higher energy orbit. An X-ray is emitted when the electron falls to a lower orbit.

Absorption is the transformation of the radiation energy into some other form of energy, such as heat.

Quantities Used to Describe Radiation Interactions

The probability that a reaction will occur varies with the type of EMR or particle, the medium it travels through, and the energy of the EMR or particle. These are defined by the microscopic cross section, σ and the macroscopic cross section, Σ . If a wavicle (wave or particle) is incident upon the nucleus within the cross section, the probability of interaction is one, otherwise it is zero. The standard unit is the barn/atom (or cm^2 per atom) where

1 barn (b) = 10^{-24} cm squared   9-28

The barn was devised as a unit because one scientist said something like, "compared to the typical cross section, this nucleus is as big as a barn."

Alpha Particle Interactions

Typically alpha particles are ejected at speeds of 1/20th c and energies of 4-9 MeV. They lose energy by either direct collision with an electron or electrostatic field interaction between the alpha particle and the orbital electrons. Excitation usually occurs at the end of the alpha particles range. Figure 9.5 shows the SI versus range for an alpha particle. Figure 9.6 shows the intensity versus distance. These figures (9.5 through 9.8) were drawn free hand and are not accurate, but they give an idea of how the variables change.

Figure 9.5 SI vs. distance for an Alpha particle (an approximation)

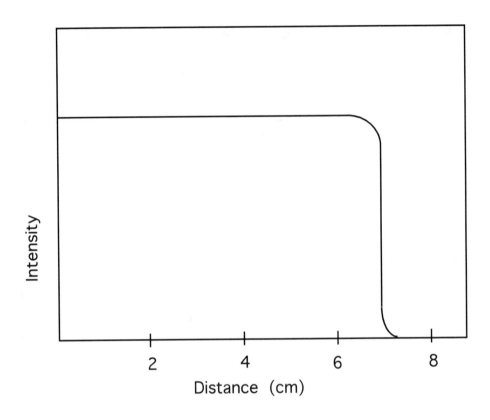

Figure 9.6 Intensity vs. distance for an Alpha particle (an approximation)

Beta Particle Interactions

Unlike alpha particles, betas are not emitted in discrete energies, but have a continuous spectrum as in figure 9.7. The beta intensity versus range is shown in figure 9.8--compare it with the alpha particle! The most frequent method of losing energy is loss by collision, leading to ionization and excitation. But, for energetic betas of >1 MeV, the beta can interact with the electric field of the nucleus and emit X-rays by what's known as Bremsstrahlung radiation (braking radiation in German). When a charged particle is accelerated or decelerated in an electric field, radiation is given off. As the electron passes close to the nucleus, it is attracted and accelerated, producing radiation. The radiation is proportional to $Z/(m^2)$. The amount of Bremsstrahlung radiation is negligible, except for the electron, since its mass, m, is so small. (Radiation varies as $1/m^2$, the lower the m, the higher the radiation.)

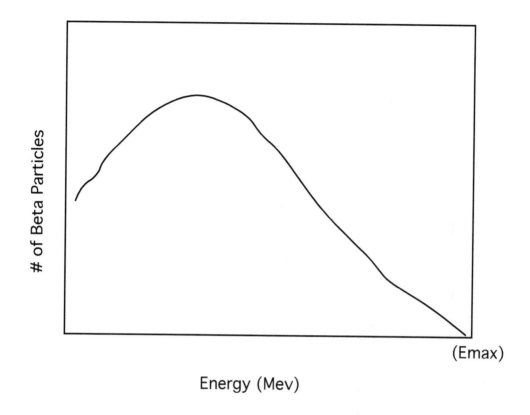

Figure 9.7 Beta particle spectrum (an approximation)

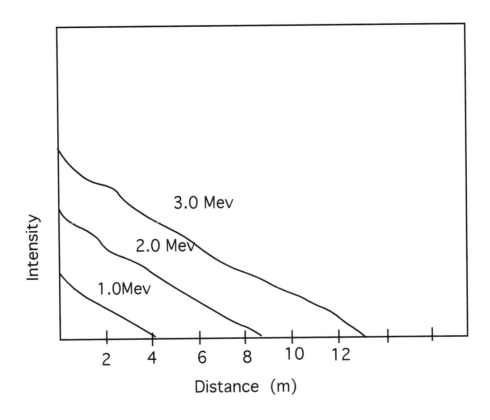

Figure 9.8 SI vs. distance for a Beta particle (an approximation)

The stopping power for electrons decreases as Z increases because materials with high Z have fewer electrons per unit mass. As Z increases however, the multiple scattering of electrons increases. These two effects tend to cancel out, so that density of a material is a good measure of its stopping power. However, the higher the Z, the more Bremsstrahlung radiation is given off, so most shields have low Z.

Gamma Interactions

There are three processes, the photoelectric effect, the Compton effect (or Compton scattering) and pair production.

Photoelectric Effect

The gamma ray is absorbed by an inner shell electron which is ejected from the atom. This electron ionizes and excites the medium. The other orbital electrons fall to lower orbits and emit X-rays.

Compton Effect

The gamma ray hits an unbound electron, which is scattered by the gamma ray and changes the angle of the gamma ray. As the energy of a gamma ray increases, bound electrons behave as free electrons. Conservation of kinetic energy and momentum are maintained. Thus, the photon (gamma ray) loses energy (becomes longer in wavelength) and changes its path; it is scattered at an angle theta.

Pair Production

When a gamma has at least 1.022 MeV of energy (two electron masses), then it can interact with an electrostatic field and form an electron/positron pair. Since the nucleus has a greater electric charge, most pair production reactions involve the nucleus.

Neutrons

Neutrons can be classified by their energy. Thermal neutrons are in thermal equilibrium with their surroundings. This means that when a large number of them diffuse through a non-absorbing medium there is no energy exchange. Resonance neutrons are in the energy range of 1 to 100 eV. Many nuclides have a strong absorption of these neutrons, and this is called resonance absorption. Neutrons in the range of 1 kEv to 0.5 MeV are called intermediate neutrons. Intermediate neutron behavior strongly affects the design of high performance shields. Slowing them down is where power is generated in a reactor. Fast neutrons are in the range of 0.5 to 15 MeV, and are the ones produced by nuclear fission and fusion.

Neutron Interactions

One way to look at neutron interactions are as elastic and nonelastic scatterings. Another way is by scattering versus absorption.

Elastic scattering is simply billiard ball physics. One ball hits the other, both go flying off in different directions. Both have changes in velocity.

Nonelastic scattering can be further broken down into:

Inelastic scattering (n,n'), where the neutron combines with the nucleus and a lower energy neutron, n', is emitted, and a gamma ray is emitted to get the nucleus to the ground state.

Nuclear reactions where the collision of the neutron with the nucleus changes the atomic number or mass. Nuclear reactions consists of the following:

n,2n and n,3n reactions where the nucleus emits 2 or 3 neutrons after absorbing the incident neutron. This generally occurs at about 10 MeV, although it occurs at less than that for heavy and light nuclei. Beryllium, used in space reactors, has loosely bound neutrons, and these can be easily ejected in an (n,2n) reaction.

Fission reactions where the nucleus splits into two new atoms, accompanied by the release of neutrons and gamma rays.

Radiative capture (n, γ) where the nucleus emits one or more gamma rays (equation 9-29)

$$^{235}_{92}U + ^{1}_{0}n \rightarrow ^{236}_{92}U^* + \gamma \qquad \qquad 9\text{-}29$$

Where * is an excited state.

And charged particle reactions where the nucleus emits either a proton or alpha particle. Equation 9-30 and 9-31 show some of those reactions.

$$^{3}_{2}He + ^{1}_{0}n \rightarrow ^{4}_{2}He^* \rightarrow ^{3}_{1}T + ^{1}_{1}p + 0.764\text{Mev} \qquad \qquad 9\text{-}30$$

$$^{10}_{5}B + ^{1}_{0}n \rightarrow ^{7}_{3}Li^* + ^{4}_{2}\alpha + 2.30\,Mev \qquad \qquad 9\text{-}31$$
$$^{7}_{3}Li^* \rightarrow ^{7}_{3}Li + \gamma\,(0.47\,Mev)$$

Hazards of Neutrons

The damage done to the human body is a function of energy. Neutrons, being uncharged, move through the body freely. The more radiosensitive organs in the body will be damaged by these neutrons (as they will by gamma rays, which also move freely through the body).

Fast neutrons cause many reactions (collisions) in the tissues of the body. Since we are mostly water (H_2O), 80 to 95 percent of the energy is deposited in hydrogen scattering reactions. The recoil protons lose their energy by ionization. Therefore, fast neutrons are more hazardous than thermal neutrons.

Thermal neutrons can produce damage through hydrogen scattering and absorption.

Absorption:

$$^{1}_{0}n + ^{1}_{1}H \rightarrow ^{2}_{1}D + \gamma \qquad \qquad 9\text{-}32$$

$$^{1}_{0}n + ^{14}_{7}N \rightarrow ^{14}_{6}C + ^{1}_{1}p \qquad \qquad 9\text{-}33$$

Note that the proton has a + 1 charge, so this is a charged particle moving in the body, causing more damage.

144

Since some of the neutron interactions produce heavy recoil particles, recoil protons, and gamma rays, the transfer of energy to the body tissue will vary significantly with respect to each particle or ray. The relative effectiveness, however, of neutrons in producing damage is 2.5 to 10 times that of gamma rays.

A brief synopsis on how you die from radiation

One key thing that we want to do is to keep a person's exposure to radiation "as low as reasonably achievable" (ALARA). There are two exposure limits, one for the general population, and one for occupational workers. (Astronauts have their own limits). The International Commission on Radiological Protection, ICRP, sets these limits. Three fundamental units for radiation protection have been used, the Roentgen (R), the rad (radiation absorbed dose), and the rem (roentgen equivalent man). The roentgen measures how much exposure happened (you were exposed to 200 bullets). The rad measures how much of that radiation was absorbed (only one bullet hit you). The rem measures a biological dose equivalent, in other words what is the effect of 1 rad on a person in a biological sense. Does 1 rem cause a skin sore, or does it kill the person.

When radiation was first investigated, the Roentgen was the unit of measurement. It was defined as the amount of gamma ray or X-ray radiation that yields one electrostatic unit of charge in a cubic centimeter of dry air at standard conditions, 273 K and 101.32 kPa (1 atmosphere). This is equal to 2.58 times ten to the minus 4 coulomb per kilogram of air. But the roentgen is what you are exposed to, the rad is what you absorb. One roentgen of X or gamma at 0.1 to 3.0 MeV produce a 0.87 rad dose in air and roughly 0.98 rad in soft tissue, such as skin and organs.

One rad is the amount of radiation needed so that the material absorbs 100 ergs per gram. This converts to 62.4 times ten to the sixth MeV per gram absorbed in the substance. The SI unit of absorbed dose is the gray (Gy) and is defined as the amount of radiation that will cause the absorption of one joule per kilogram. One Gy equals 100 rads.

Different types of radiation produces different effects biologically speaking. The effect depends upon the number of rads and the rate or spacial distribution at which the incident radiation loses energy in the biological tissue. This energy loss rate is known as a linear energy transfer (LET). Basically, the higher the LET, the more harmful to tissue. Radiation that is easily stopped (alpha or protons) creates more damage in tissue than does X-ray, gamma ray, or beta particles. This is particularly true of sensitive organs, such as the bone marrow, gonads, lymphoid areas, and embryonic tissues. Dose equivalent (H) has been created to address the severity of harmful effects from radiation on tissue, including the entire organism. It is defined as

$$H \equiv D \times QF \times N$$

9-34

Where D is the absorbed dose, QF is a quality factor, and N represents any other modifying factors that are required to account for biological effects of a particular radiation. Fortunately the ICRP defines N to be 1.

H is expressed in terms of rem, (roentgen equivalent man), or rem = rad x QF   9-35

or in SI units

Sievert = gray x QF; 1 Sievert = 100 rem, 1 gray = 100 rad   9-36

Radiation Protection

Radiation protection may be defined as protection of people, their children, and the population of the world from the harmful effects of a radiation, while keeping the benefits of nuclear energy. Remember that radiation therapy in medicine saves lives throughout the world. The harmful effects can be divided into two classes: somatic and genetic. A somatic effect involves the person, while a genetic involves the persons descendants. There are two other categories, stochastic and non-stochastic. Stochastic means that the probability of an effect occurs as a function of the radiation dose, with no threshold. Non-stochastic effects are effects those whose severity varies with the dose, and this may mean that there is a threshold to dose (i.e. no effect for a small dose).

The ICRP defines radiation protection is accomplished by preventing harmful non-stochastic effects and to minimize the occurrence of stochastic exposure effects. Both of these are with respect to risk levels that are generally acceptable. To prevent non-stochastic, they set the dose limits at low values. This means that the person cannot be exposed to the threshold exposure, even if the person is exposed to that level of radiation for their career (occupational worker) or their entire life (general population--us!).

Some effects are stochastic (any exposure may damage) but because of the natural radioactivity in the environment, we cannot guarantee that harm will not happen to people exposed to low levels of radiation. Thus ALARA.

When radiation hits a person, two types of effects can occur to their cells, direct and indirect. Direct effects are ionization and excitation of atoms in the cell, which is independent of the material surrounding those atoms. The indirect effect occurs when the radiation produces chemicals which interact with the cells. These include, H^+, OH^-, and H_2O_2 and other highly reactive chemicals, which do very nasty things to the cells. Both of these effects start a series of chemical reactions that are quite complex in the effects on the cells. If these reactions involve the DNA it can kill the cell.

Somatic effects cause cell injuries that interfere with the maintenance of bodily functions, for example the formation of blood. Genetic effects cause cell injuries in the reproductive organs. If you damage the germ cells, which pass along genetic characteristics to future generations, you can cause mutations. The most common somatic effect at low levels of radiation is cancer.

Radiation induced cancer in humans has been studied from various groups of people; scientists investigating radiation (tickling the dragon's tail), radium watch dial painters (paint brushes), uranium miners (inhalation), atomic bomb survivors of Hiroshima and Nagasaki (fallout), and people involved in the 1954 Bikini nuclear explosion test, amongst others (such as patients intentionally and unintentionally exposed to X-rays).

As examples, in 1917, the radium dial painters were instructed to make a fine point on the brushes with their lips before painting the dials with the radium. This caused them to ingest radium. They also painted their nails, faces, and teeth so they would glow at night!

In 1946, Dr. Louis Slotin was performing a test to determine how to make plutonium critical. He was using two hemispheres of beryllium to reflect the neutrons back into the sphere of plutonium. He didn't use the spacers to prevent the sphere from completely closing, he used a screwdriver. It slipped, the upper hemisphere fell, and the plutonium went critical. There was a flash of blue light, caused by the radiation hitting the air. Slotin pulled the upper hemisphere off the bottom and the plutonium went sub-critical. Too late, he absorbed 1000 rads of neutrons and gamma rays and died nine days later. This was the second scientist to die doing these experiments with the same plutonium sphere. In 1945 Harry Daghlian was doing an experiment using tungsten carbide bricks, and dropped one on the sphere. It went critical, he removed the brick, and died 25 days later. That plutonium was called the Demon Core.

For human beings, leukemia is the most common type of cancer caused by radiation. The number of cases increase as the dose increases for accumulated doses above 50 rem (0.5 Sv) and high dose rates of 1 rem/minute (10 mSv/min). Other types of cancers can occur, however. For instance, there are certain radioisotopes that are attracted to bones. High concentrations of these radioisotopes can cause bone cancer. That's what happened to the dial painters. The thyroid gland needs iodine, radioactive iodine, particularly I 131, is captured by the thyroid just like normal iodine. This is why during the Chernobyl accident governments told people to eat lots of iodized salt. The thyroid only absorbs as much iodine as it needs, so if the people are getting enough iodized salt, then they won't absorb the radioactive isotope. Radon is a radioactive noble gas. Since it doesn't bond with anything, it doesn't go into solution, and thus stays in the lungs. This causes lung cancer for those who inhale radon, such as the uranium miners. High level exposure to the skin for many years may cause skin cancer. Below about 10 rem (0.1 Sv), no experimental human or animal exposure data show any cancer risks. In mammals, no cancer has been identified at the 5 rem per year level. The ICRP chose that limit for maximum exposure for radiation workers. This, as you can see is quite conservative. Table 9.3 shows that the background radiation (environmental subtotal) in the United States exposes people to roughly 0.1 rem per year (1.0 mSv/yr). This doesn't include smoking or if you live in a state with Radon in the soil.

Hormesis

There is an alternate view to radiation exposure called hormesis. Hormesis is the effect that a low level of exposure to a toxic substance actually causes the body to react in a positive way. As an example, low level exposure of dioxin in rats causes a decrease in the rate of cancer. This is not what most medical personnel would assume or agree with, dioxin causes cancer. Similarly, low level exposure of gamma rays in mice shortly before the mice were exposed to high levels of gamma rays actually caused a decrease in cancer rates. Studies by the Swedish Army and others confirm this effect, yet the dose limits by the ICRP have not changed. It seems that exposure to radiation up to 50 mSv is good for humans. So, we have to ponder whether the ICRP limits are correct.

Definition of Occupational Worker

What is an occupational worker?

Title 10, Part 19, of the *Code of Federal Regulations* (10 CFR Part 19), "Notices, Instructions, and Reports to Workers: Inspection and Investigations," provides instructions to radiation workers. Specifically, 10 CFR Part 19 requires that all individuals who, in the course of their employment, are likely to receive a dose of more than 100 millirem in a year, must receive adequate training to protect themselves against radiation. Also, these individuals have the right to know the amount of radiation to which they have been exposed. In addition, radiation workers have the right to ask the NRC to conduct an inspection if they believe their working environment has safety problems.

Tables 9.1 through 9.8 show LET, QF, and dosages.

Table 9.1 Linear Energy Transfer factor versus quality factor.

Quality Factor (QF)	Average LET in Water (LET) (keV/micrometer)
1	<3.5
2	7
5	23
10	53
20	175

Table 9.2 Recommended quality factor values.

QF	Type of Radiation
1	X-rays, gamma rays, electrons (beta -), positrons (beta +)
10	Neutrons (fast), protons, and singly charged particles of rest mass greater than 1 amu of unknown energy
20	Alpha particles, multiply charged particles, and particles of unknown charge and/or unknown energy.
	(When neutron energy spectrum is known)
2	thermal neutrons
2.5	0.01 MeV neutrons
7.5	0.1 MeV neutrons
11	0.5 MeV neutrons
11	1.0 MeV neutrons
6.5	10 MeV neutrons

Table 9.3 Summary of estimates of annual average whole-body dose rates in the US (1970).

Radiation Source	Average Dose Rate mrem/year [mSv/year]
Environmental	
1. Natural	
a. External	
Cosmic Rays	28 [0.28]
Terrestrial	50 [0.50]
b. Internal	22 [0.22]
Radioisotopes In Body (natural)	
2. Global Fallout	4 [0.04]
3. Nuclear Power Industry	0.003 [3 x 10^{-5}]
Environmental subtotal	~104 [~1.04]
Medical and Therapeutic	
1. Diagnostic	72 [0.72]
2. Radiopharmaceuticals	1 [0.01]
Medical subtotal	~73 [~0.73]
Occupational Radiation Worker	0.8 [0.008]
Miscellaneous	2 [0.02]
TOTAL	~180 [~1.8]

Table 9.4 Acute radiation effects from whole-body exposure to gamma radiation.

Acute Irradiation Level rem [sievert]	Acute Somatic Effect
15-25 [0.15-0.25]	Subtle white blood cell reductions; not generally apparent from exposure to one individual unless a blood sample was taken before as well as after exposure
50 [0.5]	Reduction in white blood cell count (WBC) after exposure; the count returns to normal in a few weeks
75 [0.75]	10% chance of nausea
100 [1.0]	10% chance of temporary loss of hair
200 [2.0]	90% chance of radiation sickness; moderate depression of WBC fractions
400-500 [4-5]	50% chance of death within 30 days without extensive medical treatment
>600 [>6]	Lethal to most people in 3 to 30 days; even with extensive medical treatment, death is likely within a few months from infection and hemorrhage
>10,000 [>100]	Lethal within 24 hours from central nervous system damage.

Table 9.5 Effects to the body from localized exposure to X and gamma radiation.

Organ of Body	Acute Irradiation Level rem [sievert]	Acute and Delayed Biological Effects
Skin	300 [3.0] 1500 [15] 5000-7000 [50-70]	Erythema or sunburn effect noticeable Raw, moist skin surface where irradiated Ulceration, slow healing, possible skin cancer
Gonads	50 [0.5] 250 [2.5] 600 [6.0]	Brief functional sterility in males only Sterility for 1 to 2 years in both male and female Permanent sterility
Eye	200 [2.0] 600 [6.0]	Change in optic lens opacity Clinically significant cataract
Fetus	10-20 [0.1-0.2]	Significant probability of malformation, if irradiation occurs in first 3 months of pregnancy

Table 9.6 Occupational worker permissible exposure levels.

Type of Radiation Exposure	Exposure period	Maximum Possible Dose Equivalent rem [mSv]
Whole-body, head and trunk, gonads, lens of Eye, red bone marrow, active blood forming organs	Year Calendar quarter (13 weeks)	5 [50] 3 [30]
Unlimited areas of the skin, except hands and forearms; other organs, tissues, and organ systems, except bone	Year Calendar quarter	15 [150] 5 [50]
Bones and forearms	Year Calendar quarter	30 [300] 10 [100]
Hands and feet	Year Calendar quarter	75 [750] 25 [250]

Table 9.7 permissible exposure levels to the general public.

Type of radiation exposure	Annual Max permissible dose equivalent		Annual Max permissible dose equivalent	
	Based on Dose to individuals at points of maximum exposure		Based on average Dose to a suitable sample of the exposed population	
	(rem)	(mSv)	(rem)	(mSv)
Whole-body, gonads, or bone marrow	0.5	5	0.17	1.7
Other organs	1.5	15	0.50	5

Table 9.8 Radiation exposure limits recommended for space-flight crew-members.

Ancillary Reference Risks

Constraint	Primary Reference Risk* rem [Sv]	Bone Marrow** (Dose Equivalent at 5 cm) rem [Sv]	Skin (Dose Equivalent at 0.1 mm) rem [Sv]	Ocular Lens (Dose Equivalent at 3 mm) rem [Sv]	Testes (Dose Equivalent at 3 cm) rem [Sv]
1 year average daily dose		0.2 [0.002]	0.6 [0.006]	0.3 [0.003]	0.1 [0.001]
30 day maximum		25 [0.25]	75 [0.75]	37 [0.37]	13 [0.13]
Quarterly maximum***		34 [0.34]	105 [1.05]	52 [0.52]	18 [0.18]
Yearly maximum		75 [0.75]	225 [2.25]	112 [1.12]	38 [0.38]
Career limit	400 [4.0]	400 [4.0]	1200 [12.0]	600 [6.0]	200 [2.0]

* Dose equivalent at 5 cm depth in tissue
** Whole body exposure
*** May be permitted for two consecutive quarters followed by 6 months of restriction from further exposure to maintain yearly limit and includes all occupational exposures.

Problems

1. Martian scientists are planning to send a probe to the dead planet one closer to the sun from them. Their spacecraft has an RTG fueled by Marsium-242, which was in- vented on Mars. Marsium decays to Terrium by a Beta minus decay. Terrium decays to Venusium with an Alpha decay.

$$^{242}_{110}Mr \xrightarrow{\beta^-} {}^A_B Tr \xrightarrow{\alpha} {}^C_D Vs$$

The half life of Mr is 4 years, the half life of Tr is 11 years.

a) what are the numbers A, B, C, and D?

The scientists measured the specific power of Mr as 10 watts per gram, and the Tr as 5 Watts per gram. The spacecraft is fueled with 10 grams of pure Mr.

b) After two years of flight to the dead planet, what is the TOTAL heat generation of the RTG?   (both the Mr and Tr)

2. Carbon-14 ($^{14}_6C$) dating of organic materials is based on the fact that the organisms is absorbing Carbon-14 from the Earth's atmosphere. The Carbon-14 is constantly decay- ing, but is replaced by the organism. This stops at death, and the Carbon-14 decays without being replaced. The radioactive carbon amounts to about 0.10 of the total carbon in the organism. Wood in an excavation yields a Carbon-14 amount of 0.080. The half life of Carbon-14 is 5730 years, what is the age of the wood?

3. $^{218}_{84}Po$ emits either an alpha particle or a beta + or beta - particle.
a) Write all the possible equations for the reactions.
b) The half life for the alpha is 3.05 minutes, what is the decay constant?
c) Determine the number of atoms in a sample with an activity of 100 Micro-Curies.
d) c) but with an activity of 100 Becquerels.
e) for c) calculate the activity in microCuries after 1, 2 and 6 half lives.

4. At the Fukushima Daiichi nuclear plant on 21 March 2010 the radiation was reported to be 400 milliSieverts per hour. How long, in minutes, could a Nuclear worker work in that environment and not exceed the quarterly dose, assuming the radiation was gamma radiation?

For problems 5 and 6 use the whole body column in the charts

5. An asteroid miner comes across an object of Extra Terrestrial origin. Knowing there is a lot of money in this, the miner decides to investigate. He sends a remote

probe which reports that the object is radiating gamma rays at 500 mrad/hour, fast neutrons at 300 mrads/hr, and alpha particles at 400 mrads/hour. The miner dons his spacesuit and investigates the object further for 3 hours.

For the following a) and b) we will say he is an astronaut.

a) What is the dose equivalent (in REM) for each type of radiation that he absorbs and what is the total dose equivalent experienced by the miner?

b) Ignoring the alpha particles, how long can the miner stay outside investigating the object and not exceed the quarterly dose?

6. The energy from sunlight incident at the Earth's surface is approximately 0.1 Watts/cm^2. Assuming the average body has an area of one square meter, that the person stays out in the sun for 5 hours, that the person weighs 60 kg, and that the energy of the sunlight is as bad as gamma rays (i.e. it causes radiation damage as gamma rays would):

a) How many Sv does the person absorb? (Assume that the shielding caused by the bathing suit is negligible, which, by today's standards [for either sex] is a good assumption).

b) What kind of health problems would this person experience? In other words, is this person going to be sick, lose their hair, etc. Be brief, 10 words or less.

7. A nuclear weapon's yield is expressed in Kilotons or Megatons of TNT. A kiloton of TNT is 4.18×10^{12} Joules. In one Star trek episode Kirk drew off 4 ounces (let's call it 0.5 kg) of antimatter to destroy a creature. If all of that antimatter reacts with matter and converts into energy, how many kilotons of TNT was produced? (Compare to Hiroshima, which was 13 kilotons, or Nagasaki, 22 kilotons). Use E=mc^2.

Chapter 10, Nuclear Reactors

All nuclear reactors have three basic applications.

1. Intense neutron beams/fields for scientific experiments.

2. Production of new materials by neutron irradiation.

3. Liberate thermal energy for power generation.

Before we get into the workings of a reactor, first some definitions.

A **fissile nuclide** is a nuclide (U 233, U 235, or Pu 239) that will split upon absorbing neutrons of any energy.

A **fertile nuclide** is a nuclide that can be changed into a fissile nuclide by absorbing a neutron (Th 232, and U 238)

$$_{90}^{232}\text{Th} + _{0}^{1}\text{n} \longrightarrow _{90}^{233}\text{Th} \xrightarrow{\ _{-1}\beta\ \text{decay (T1/2} = 21.38\ \text{minutes}} _{91}^{233}\text{Pa} \longrightarrow _{92}^{233}\text{U}$$

$$_{-1}\beta\ \text{decay (T1/2} = 29.967\ \text{days}$$

$$_{92}^{238}\text{U} + _{0}^{1}\text{n} \longrightarrow _{92}^{239}\text{U} \xrightarrow{\ _{-1}\beta\ \text{decay (T1/2} = 23\ \text{minutes}} _{93}^{239}\text{Np} \longrightarrow _{94}^{239}\text{Pu}$$

$$_{-1}\beta\ \text{decay (T1/2} = 2.3\ \text{days}$$

A **chain reaction** is a reaction that is repetitive. For fission this means that a fissile nucleus gets hit by a neutron, absorbing its energy, splits, and releases more neutrons.

There are usually two smaller and less massive elements that are produced by this splitting. These are called fission products.

We can now define a nuclear reactor as a device containing fissile material such that a chain reaction can be sustained and controlled. Controlled is a very important concept.

A Thermal reactor is one where thermal neutrons (neutrons in thermal equilibrium with their surroundings) are the main cause of fission, while a fast reactor uses fast neutrons, i.e. >100 keV.

The multiplication factor, K is defined as

$$K \equiv \frac{\text{\# of nuclear fissions (or neutrons) in one generation}}{\text{\# of nuclear fissions (or neutrons) in the immediately preceding generation}}$$

When K = 1, then the reaction is critical, that is, it is self sustaining. It will have a constant power output, the reactor doesn't have to be adjusted. If K < 1, then the reaction is subcritical, and the fissions occurring per time unit will eventually reach zero. If K > 1, the reaction is supercritical and the fissions per time unit will increase, and the power out will go up. As K goes up or down, the number of neutrons goes up or down. This is called in the literature as the Neutron population.

The generic space nuclear reactor

Figure 10.1 shows a generic space reactor, which consists of:

1. the core
2. nuclear fuel
3. coolant
4. a reflector
5. control drums
6. radiation shield
7. a moderator (thermal reactors only)

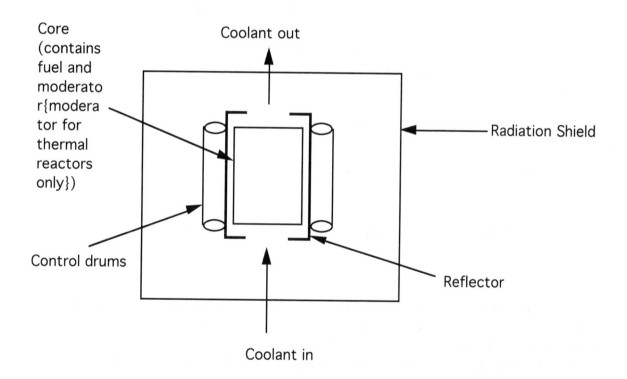

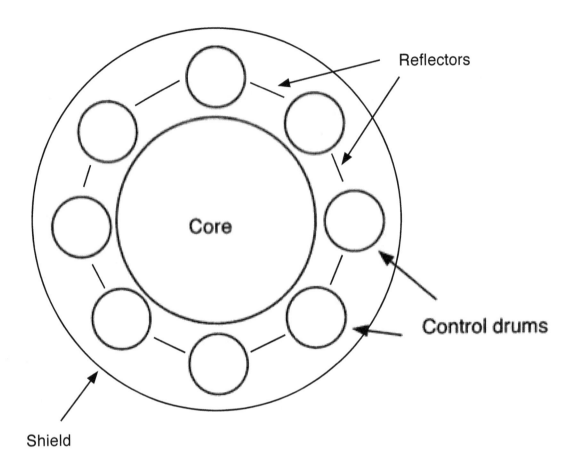

Figure 10.1 Generic Space Nuclear Reactor, side and top views

The core contains the fuel, the moderator, structural materials, and coolant passages (such as heat pipes). The nuclear fuel, such as enriched (i.e. 93.5%) U 235 can be in solid, liquid, or gaseous form. The moderator is a low mass material, such as hydrogen, that slows down (moderates) neutrons from their fission energy to thermal energies (usually in about 20-30 collisions).

The coolant can be a pumped loop system with liquid metal or a gaseous xenon-helium mixture, or can be a self contained heat pipe system. In fast reactors, the coolant must be carefully chosen in order to not slow down the neutrons (we don't use water!!)

The reflector, typically beryllium, is located next to the core and scatters the neutrons back into the reactor to reduce neutron leakage. It also flattens the neutron flux in the core, i.e. it keeps the core and coolant uniformly hot.

Control drums or rods are made of a neutron absorber, such as boron (boron carbide). Movement of these rods or drums adjusts k. Drums have 1/2 of the cylinder an absorber, the other half a reflector, typically Beryllium. The drums are rotated with the absorber towards the core, absorbing neutrons, or rotated with the reflector towards the core, increasing the neutrons used for a reaction. Rods are pushed into the core to absorb neutrons.

The radiation shield has to shield neutrons and gamma rays from the essential stuff, such as the payload (the satellite or science instrument) and the astronauts. Lithium Hydroxide (LiH) is used to shield neutrons, Tungsten (W) is used to shield gamma rays.

The other way to shield is to keep the source farther away. Watch the movie, 2001 a space odyssey. You will see that the ship has the reactor near the engines, on a truss that is long. All the fuel is on the truss, and the living quarters are well away at the other end of the truss. You can also make a shield that is called a two pi shield, which is basically a hemispherical shield that is pointed towards the payload. Therefore the payload is shielded, but what's behind the reactor isn't. This saves mass versus a four pi shield, which is spherical.

Thermal versus Fast reactors

Thermal

The majority of terrestrial reactors are thermal. They are the easiest to make critical, and they also require less fissile material. Example: For Uranium 235 the cross sections for thermal and fast neutrons are:

σ_n (0.025 eV) = 577 barns
σ_n (2 MeV)   = 1.3 barns

The higher cross section makes it easier to make the reactor critical.

Fast

There is no moderator, and the coolant has to have a high A, Sodium-Potassium (NaK) or Sodium (Na) are coolants of choice.

These reactors require a large amount of fissile material. The Fermi reactor in Chicago (built in the 1940's) was a fast reactor using 444 kg of U 235. A comparable thermal would use 20 kg.

U 235 and Pu 239 can be used as fissile material.

There are a few fast reactors in the world, mainly for research, but some supplying power. The United States built several, but the have all be shut down.

So, why build fast reactors if they take so much fuel? The average number of prompt and delayed neutrons released per fission, ν, is more favorable for breeding new fissile fuels. Hence, these are called breeder reactors. For U 235 the number varies from 2.4 to almost 3 with the energy of the neutrons varying from 0 to 4 MeV. For Pu 239 the number varies from almost 2.9 at about 10^2 KeV to about 3.2 at about 10^3 KeV. You put U 238 or Th 232 in the blanket around the core which supplies fast neutrons. The fast neutrons make more fissile fuel than is consumed. Think about that, this is a real renewable source of energy!

The fission process

Nuclear fission starts with the absorption of a neutron by the nucleus. The nucleus is then in an excited state. If it does not have enough energy, called the threshold energy, an activation gamma, (n, γ) ray will be emitted. If enough energy was absorbed, the nucleus will fission, releasing prompt neutrons, prompt gamma rays, and fission fragments. (The definition of prompt is at the time of fission). Table 10.1 shows calculated values of the fission threshold energy by mass number. Only very heavy nuclides (A > 230) have reasonably low threshold energies. Table 10.2 shows fission threshold for selected heavy nuclides. If the energy is less than zero, then a neutron with essentially zero kinetic energy can cause the nuclides to undergo thermal fission. All the other nuclides can only undergo fast fission.

Table 10.1 Neutron fission thresholds as a function of Mass.

Mass Number A	Fission Threshold (MeV)
16	18.5
60	48
100	47
140	62
200	40
236	approximately 5

Table 10.2 Neutron fission thresholds of heavy nuclides.

Target Nucleus	Compound Nucleus	Fission Threshold (MeV)
232 Th	233 Th	1.3
233 U	234 U	<0
234 U	235 U	0.4
235 U	236 U	<0
236 U	237 U	0.8
238 U	239 U	1.2
237 Np	238 Np	0.4
239 Pu	240 Pu	<0

The fission almost always produces two fission fragments, although more than that is possible. (It is such a rare event that it can be ignored). The nucleons are bound tighter within the fission fragments than they were in the original nucleus. Therefore, a very large amount of energy, typically 200 MeV, is released (Table 10.3). The fission also results in the release of more neutrons. The average number released, ν, is a function of the neutron energy and the type of fissile nuclide (Uranium, Plutonium, etc.). Thermal fission of U 235 produces 2.5 neutrons per fission (on average), while for Pu 239 it is 2.9.

Table 10.3 Typical energy distribution for Uranium 235 fission.

Energy Form	Energy Released (MeV)	Energy Potentially Recoverable (MeV)
Kinetic Energy of Fission Fragments	168	168

Decay of Fission Products
  Beta Radiation
  Gamma Radiation
  Neutrinos
	8
	7
	12
		8
		7
		0

Prompt (Fission) Gamma Radiation	7	7
Kinetic Energy of Fission Neutrons	5	5
Capture Gamma Radiation	0	3-12
Totals	207	198-207

Some of the fission fragments also possess more neutrons than "normal" atoms need and are unstable. They typically have half lives of about one minute and emit delayed neutrons, delayed beta and gamma radiations in their attempt to become stable. The delayed neutron fractions for U 235 and Pu 239 are about 0.7 and 0.2 percent, respectively.

The fission yield is the percent of the total number of fissions that produce fission products of a certain mass number. Generally low and high mass fission products have low probability (0.1 to 0.01 %) and the highest probability exists for almost equal masses being produced after fission. As the energy of the initial neutron increases, this probability becomes even more pronounced.

The factor, η, is defined as the number of neutrons released by fission per neutron absorbed in the fissile fuel. Since radiative capture (n, γ) occurs more than the fission process for neutrons, η is always less than ν.

Finally, there are spontaneous fission neutrons from transuranics produced in the core. Also you can have decay alpha, beta, and gammas from the fission products, activation products, and transuranic decay (where the half life varies from seconds to millions of years).

After-heat

This is also called decay heat and it is the thermal energy released by the radioactive decay of the core (mostly fission products) after the chain reaction has stopped. After a reactor has operated for a couple of days, the decay of all the fission products, generate 10 percent of the total thermal power output of the system. These nuclides are primarily beta and gamma emitters. This power output, P_s, is a function of the number of fissile products that have been produced. This is a function of the power level that the reactor was operated at just before it was shut down, P_o, and the time, t_o, that it operated at P_o.

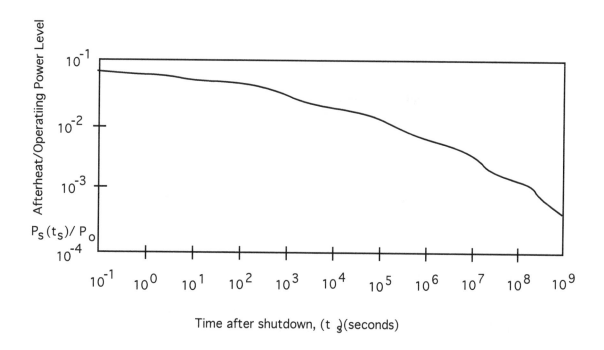

Figure 10.2 After-heat (an approximation)

For space nuclear reactors after-heat forces a to design of the reactor to use auxiliary power supply to keep the coolant moving through the core, or to use heat pipes to remove the heat from the core.

Reactor Control

Control drums and rods are used to change the reactivity of the reactor to:

1. Change the power level (up and down)

2. Adjust the reactivity changes that occur over time caused by the burn-up of fuel in the core and by fission product buildup, some of which absorb neutrons

3. Compensate for temperature induced reactivity changes

4. To allow rapid shutdown in emergencies (SCRAM, which does NOT stand for Safety Control Rod Axe Man, that is a myth)

The reactor's temperature significantly affects criticality. The temperature coefficient of reactivity, α_T , is defined as

$$\alpha_T \equiv \frac{d\rho}{dT}$$

10-1

where $d\rho$ is the reactivity change ($ or cents--100 cents to a $)
and dT is the temperature change in Kelvin

If k is close to unity, then this becomes

$$\alpha_T \approx \frac{1}{k}\frac{dk}{dT}$$

10-2

This is also called alpha prompt because the fuel temperature reacts faster to changes in power level. This value establishes the reactor's response to changes in temperature or power level. For safety purpose, reactors should have a negative value of alpha prompt. This means that dk/dT is negative. So as T increases, k decreases, the power level goes down, and T drops. The reactor is stable with respect to temperature induced reactivity changes. What actually happens is that when the temperature of the reactor increases, the U 238 cross section increases, thus the U 238 absorbs more neutrons, so less are available to continue the fission process. Thus k decreases, the reactor drops in power, and the temperature goes down. If it goes down too far, the cross section of U 238 decreases, more neutrons are available to split the atoms, and the power level increases. In the Chernobyl reactor, which had graphite as a moderator and a water cooling loop which also moderated the neutrons, there was an unfortunate design flaw. The water was effectively a neutron absorber. When the water boiled, the number of neutrons available for fission increased. Oops!

The buildup of fission products which absorb neutrons (and to some extent all of them do) produces a long term decrease in the reactivity. The two big problems are Xe 135 and Sm 149 (figures 10.3 and 10.4). The former is the most significant poison because it has a 2.65 mega-barn cross section for thermal neutrons. Because it and its precursors have short half lives, the concentration quickly reaches equilibrium in the reactor, and essentially disappears once the reactor is shut down. Sm 149 is stable and has a 74 kilo-barn cross section for neutrons. It disappears only when it absorbs a neutron. Due to its short-lived precursors, it achieves equilibrium within a few day of operation of the reactor. Its concentration stays the same after shutdown.

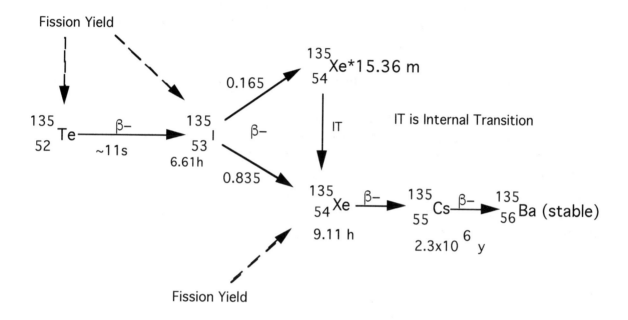

Figure 10.3 Xe 135 absorption

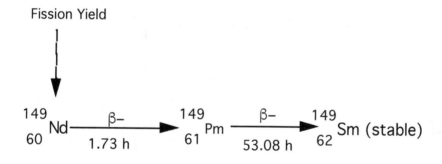

Figure 10.4 Sm 149 absorption

Because of their small size (the SNAP-10A was 3.5 meters long with a base of 1.5 meters and weighed 436 kg), the reactivity for space reactors is controlled using reflector manipulation. Reflector control mechanisms can be shutters or windows, which merely create holes in the reflector and let the neutrons leak out into space, or cylindrical drums.

Burnable poisons can also be used to control the level of reactivity. Burnable poisons are a material that has a large neutron absorption cross section, but once it absorbs the neutron, it has a small absorption cross section. They are therefore depleted during the reactors lifetime, which increases core reactivity. This compensates for the depletion of fissile material in the core and for the buildup of the fission products that absorb neutrons. Most materials have small cross sections for fast neutrons, so this method is only used in thermal reactors. For thermal reactors, this method can compensate for the Sm 149 and Xe 135 described above. Finally, as the size of a space reactor increases, reflector control gives way to control rods. This is because the volume increases and then control rods are needed.

SNAP (Systems for Nuclear Auxiliary Power) was NASA's program to develop both RTGs and Reactors. Only one US reactor has flown in space, the SNAP 10A (figure 10.5). It was a ZrH-U 235 powered reactor with control drums. Coolant was NaK-78 with thermo-electric power conversion using SiGe semiconductors. It had an output of 533 watts electric with input of 43.8 kW, or 1.22% efficiency in converting heat into electricity. It was launched into a 1328 by 1295 km orbit on 3 April 1965 on an Atlas Agena vehicle. After six days at full power, the automatic control system was shut down and the reactor was controlled by its inherent negative temperature coefficient of reactivity. After 43 days the reactor shut itself down. This was theorized to be a failure in the voltage regulator in the Agena vehicle. At that altitude, it's orbit will decay and it will enter the Earth's atmosphere in 4000 years. By then all the nasty byproducts will have decayed. Calculations show that after 15 years the fission products would produce 100 curies, after 100 years less than 0.1 curies.

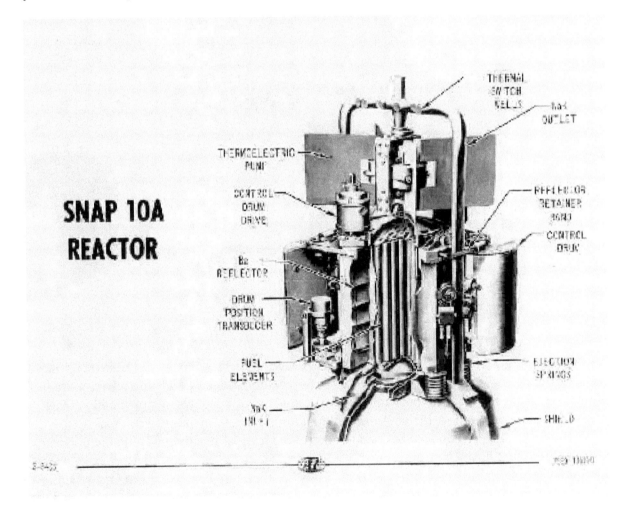

Figure 10.5 SNAP 10A Reactor Courtesy NASA

170

In 2009, Leslie Dewan and Mark Massie looked at 50 year old articles in the MIT library. They looked at molten salt reactors. They are safer than the terrestrial reactors we use today. They use liquid uranium fuel, reducing the chance of meltdown. These reactors can burn up to 96% of its fuel, only 4% is used by light water reactors. It can produce 75 times the electricity per ton of uranium. It can use spent fuel rods for its fuel, thus getting rid of them safely. The fuel is heated to 500 C, and Xenon and Krypton are off gassed, they don't stay in the fuel like in fuel rods. Remember Xenon is a big poison for reactions, and will shut the reactors down. They plan on using the reactor to heat water, turn into steam, and make power with turbines. Whether this can be used for space missions remains to be seen.

Safety

In a previous chapter, we talked about the aerospace safety requirements for RTGs. The aerospace safety requirements for reactors are slightly different. The safety requirements are to maintain sub-criticality under any and all conditions, in the Earth's atmosphere or on the Earth's surface. After the reactor has been producing power in space, the reactor will not enter the terrestrial biosphere until the fission products do not present a radiological risk. This means that the reactors will be in an orbit in excess of three hundred years.

The Soviet Union (before the breakup when they became Russia) boosted their reactors to a higher orbit. If that failed their aerospace nuclear safety policy included a core dispersal mode so that the core products are spread over a large area. The reactors are reportedly equipped with a backup system that will disperse the core in such a way as to not contaminate the people living in the dispersed area more than 0.5 rem (5 mSv) during the first year.

On 24 January 1978 a Soviet satellite called Cosmos 954 entered the Earth's atmosphere and hit Canada's Northwest Territories. People searched 124,000 sq. km. and many fragments were found. The reactor fuel was scattered over 100,000 sq. km. About 20% (4 kg) of fuel had come back to earth. The fuel varied in size from 1 mm to 0.2 mm. The effects on the debris identified on the natural environment was insignificant, according to studies made by the U.N.

Problems

1. Discuss how fast reactors have been used for research and power.

2. How many nuclear reactors has the United States launched? What were they, as in SNAP- 99?

Chapter 11, Tethers

A tether is a wire, or rope, or wire rope, that connects one spacecraft to another object. If you put a conducting tether many kilometers from a spacecraft in orbit about a planet with a magnetic field, interesting things happen. For one, the solar wind issues electrons, which travel in spiral paths around the magnetic field lines of the planet. As the conducting tether crosses the field lines, it induces an electric field, the same principle that make terrestrial generators work. Electrons that are trapped in the magnetic field lines flow down the wire. If you put an electron gun on the spacecraft to shoot electrons back into space, you complete the electrical circuit and current flows. You then have a current flowing in a wire going thru a magnetic field, so you get a force induced that opposes the velocity of the spacecraft $(\vec{F} = \vec{I} \times \vec{B})$. Conversely, if you reverse the direction of current, the velocity of the spacecraft increases. (Figure 11.1) The Electro Motive Force (emf) for LEO is a few tenths of a volt per meter. A ten kilometer tether would produce several kilovolts leading to several kilowatts of power. The TSS-1 (Tethered Satellite System 1) tried to test this, but had problems reeling out the tether. TSS-1R (R for re-flight) flew very successfully, generating a modest amount of power before the tether severed. The original plan was to reel out 20 + km to generate approximately 5000 volts. At 6 km (3.8 miles) the tether generated 195 milliamps--97 times greater than the level obtained during the TSS-1 mission. Tether voltages as high as 3500 volts and current levels of 480 milliamps were obtained. This was nearly 200 times more than TSS-1, which was limited to 60 volts due to shorter cable length. This was also better than was predicted by the analytical models for TSS-1R. The maximum length the tether was deployed was 20.27 km or 12.2 miles, it was planned for 20.59 km or 12.8 miles. Figure 11.2 is the actual tether used on TSS-1 and TSS-1R. Figure 11.3 is a photo of the boom, tether, and TSS-1R satellite.

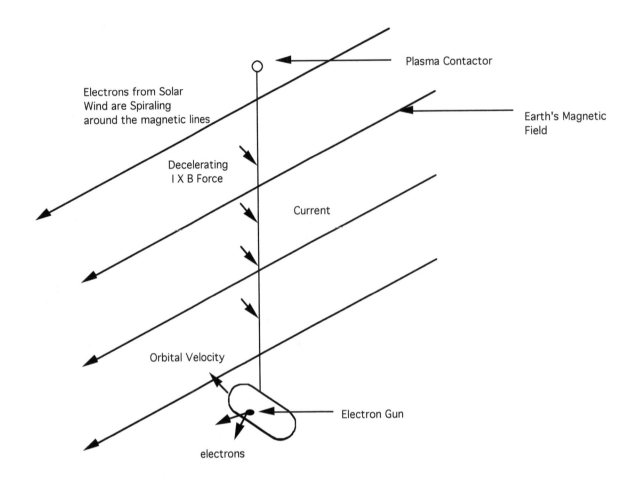

Figure 11.1 Tether Power

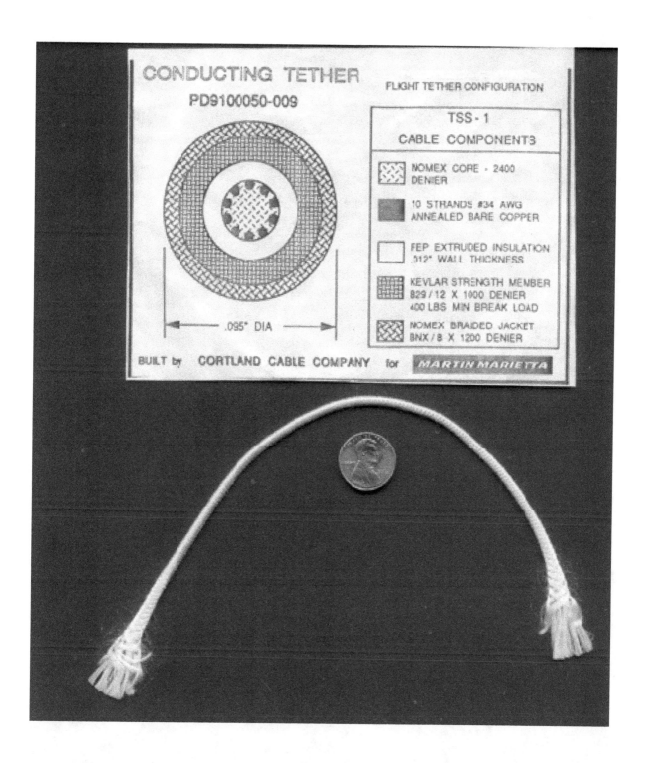

Figure 11.2 The tether from TSS-1.

Figure 11.3 TSS-1R Satellite with tether. (Courtesy NASA)

Chapter 12, Case Study-The International Space Station

Now that we have learned about various space power systems, let us look at a practical example, the International Space Station.

The International Space Station started out, under the Reagan Presidential Administration, as Space Station Freedom. The physical layout, voltages, etc. are given below.

Figure 12.1 Space Station Freedom, courtesy Rockwell International and Boeing

SPACE STATION
ELECTRIC POWER SYSTEM

Under direction of NASA Lewis Research Center, Cleveland, Ohio, the Rocketdyne team—Rocketdyne, Ford, Garrett, General Dynamics, Harris, Lockheed and Sundstrand—has completed the Phase B Electric Power System design for NASA's Space Station and Free Flying elements. This is a "sunlight to user" solar power system with power generation, management and distribution throughout the station and platforms.

NASA's phased approach features 75 kW Photovoltaic Power generation in phase 1 and adds 50 kW of Solar Dynamic Power in phase 2, giving the station a total of 125 kW. Solar Dynamic Power generation is three times more efficient than Photovoltaic, and will be used to increase the station power level from 125 kW incrementally to 325 kW. This technology represents a cost saving of $3 billion over the 30 year life of the station, when compared to a total photovoltaic configuration. The station will be in a low earth orbit between 180-250 miles above the earth, orbiting about every 90 minutes with approximately 60 minutes of sunlight and 30 minutes of eclipse. Stored energy will be used to provide power during eclipse. A fully automated computerized power management and distribution system will be used to control the generation and distribution of electricity throughout the station. There are two unmanned free-flying platforms. Phase 1 has a polar sun synchronous orbiting platform and Phase 2 adds a co-orbiting platform.

NASA expects to award hardware phase contracts to begin building the station in late 1987. First launch is scheduled early in 1994, with completion scheduled for 1996 for phase 1, and 1998 for phase 2.

CHARACTERISTICS

Station Structure
Truss Section..16 Ft. sq.
Overall Length (Phase 1) ....476 Ft.
Overall Length (Phase 2) ....673 Ft.
Overall Height (Phase 2) ....344 Ft.

Station Electric Power
Available to Users (Phase 1)..................75 kW
Available to Users (Phase 2).................125 kW
Photovoltaic (Phase 1)..........................75 kW
Solar Dynamic (Phase 2)........................50 kW
Voltage*......................440 AC Single Phase Distribution
208 AC To Load Converters
Frequency*..20 kHz

Energy Storage
Photovoltaic..........................Ni-H₂ Batteries
Solar Dynamic................Thermal Storage Salts
(Latent Heat of Fusion)

Platform Electric Power
Available to Users
Polar Platform (Phase 1) ....3.8 kW (Photovoltaic)
Co-orbiting Platform (Phase 2) ....3 kW (Photovoltaic)
Voltage*.............208 AC Single Phase Distribution
Frequency*..20 kHz

*Standard load converters will be developed to accommodate specific user requirements.

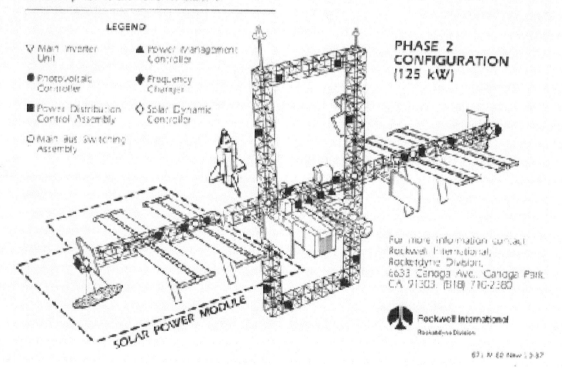

LEGEND

V Main Inverter Unit
● Photovoltaic Controller
■ Power Distribution Control Assembly
○ Main Bus Switching Assembly
▲ Power Management Controller
✦ Frequency Changer
◇ Solar Dynamic Controller

PHASE 2 CONFIGURATION (125 kW)

SOLAR POWER MODULE

For more information contact
Rockwell International,
Rocketdyne Division,
6633 Canoga Ave., Canoga Park,
CA 91303. (818) 710-2380

Rockwell International
Rocketdyne Division

Figure 12.2 Space Station Freedom electrical power, courtesy Rockwell International and Boeing

Note that there were photovoltaic solar arrays and solar dynamic power systems. The operating voltage was 440 V AC at 20 KHz. Things changed after time, here is the current International Space Station.

Figure 12.3 the International Space Station May, 2010, photo courtesy NASA. It is flying towards you.

ISS Configuration

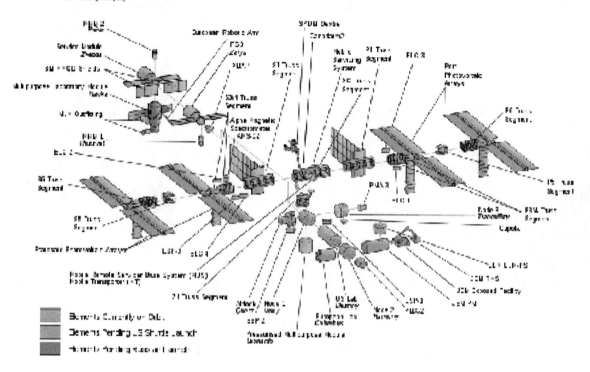

Figure 12.4 International Space Station details courtesy NASA

There are four large U.S. Photo Voltaic (PV) arrays, called the Solar Array Wings (SAW) on the Space Station. An array has four modules, each approximately 112 feet long by 39 feet wide. The arrays have a Beginning of Life (BOL) power of 31 kw and End of Life (EOL) power of 26 kw (after 15 years). The arrays rotate to face the Sun, providing maximum power to the Station. The power output is 110 kw average power, of this 46 kw average available for research. The output voltage of each array is 160 V DC. The Russian segment has solar arrays producing an additional 14 kw for research.

Each array consist of 84 panel assemblies. Two are inactive, one on the top, one on the bottom. These prevent shadowing of active panels by the containment box cover and the base. Therefore there are 82 active panels with 16,400 cells. The cells are 14.2% efficient, with an area of 8 cm² made of Silicon. Forty PV make up a module.

ISS017E012652

Figure 12.5 ISS Solar Arrays (Courtesy NASA)

There is a bypass diode every 8 cells.

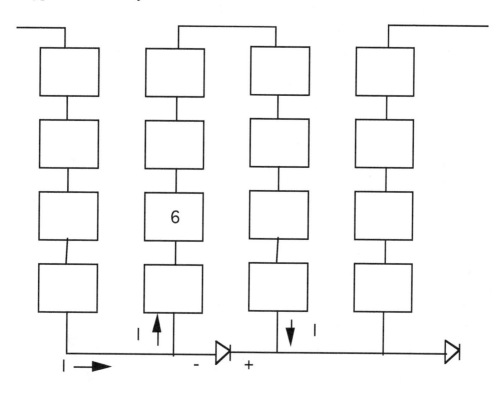

Figure 12.6 Space Station Cell Diodes

Normally the diode is biased as shown, and the current goes around the diode. If Cell 6 opens, the left side of the diode is more positive than the right side so the current flows through the diode and only 8 cells are lost.

The Station DC to DC Convert Units (DDCUs) convert the array voltage of 160 Volts down to the station voltage of 125 V DC. That's right, the Station uses 125 V DC. This means that any experiment/equipment that uses AC must convert the DC to AC. Any experiment/equipment that wants a lower voltage must have it's own DDCU of some sort.

The way the solar array power feeds the Station changed during construction. In the early missions, the Russian Zarya module supplied power to the Node 1 (Unity) and Z1 (Zenith one), which is on top of Node 1. Z1 has four gyroscopes for attitude control and a Ku band antenna for high data rate communication. The Russian to American Converter Units (RACU) change the Russian Voltage to 125 V DC. The US then launched a PV module, P6 (Port six) attached it to the top of Z1 and power was supplied to the Node via the PV array through Z1. The Laboratory module (Destiny) was launched and attached to the front of Node 1 and power came through Node 1 to the Lab. The main truss, Starboard Zero (S0) was attached to the top of the Lab. Still later the outboard trusses were installed onto S0 (Port side P1 through P5, Starboard side S1 through S6) and three PV modules (S5, S6 and P5) were added to the end of both Port and Starboard Trusses. At that point the PV modules supplied power to the Lab via the S0. Finally, the PV module on Z1 (P6) moved to the end of the port Truss, adding to the final PV array. The power routing is now complete.

In the early missions, the Shuttle would supply power from their fuel cells to the Node 1 using the Auxiliary Power Conversion Unit (APCU) which converted the 28 V DC output of the fuel cells to 125 V DC Station Power. In later missions, the Shuttle was modified so the Solar Arrays could power the Shuttle, and the Shuttle could turn off its fuel cells, allowing it to stay docked longer. This is called SSPTS, Station-Shuttle Power Transfer System.

To store the power during the shadow part of the Station's orbit, each array has 12 battery Orbit Replaceable Units (ORUs) and 6 Battery Charge/Discharge Units (BCDUs) ORUs. (ORUs is a term for anything that the astronauts can replace on orbit either in the Station or outside the Station. These are all outside the station). The heat generated by the ORUs is transferred through radiant fin heat transfer to ammonia coolant flowing in tubes on the bottom of cold plates. Ammonia cools the battery ORUs, the BCDU ORUs then electrical equipment ORUs.

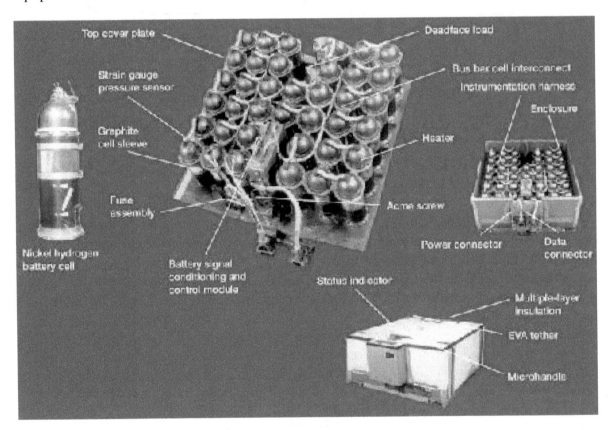

Figure 12.7 Space Station Batteries

The PV cooling has dual loops with two pumps and one radiant fin heat exchanger. The Thermal Control System (TCS) maintains a temperature of 32-50 F at the battery cell sleeve. Each SAW has a radiator to expel the heat.

Six Ni-H batteries ORUs are on one side of the solar arrays, each with one BCDU. There are another 6 batteries and another BCDU are on the other side. Each battery has a capacity of 81 A-h with a discharge voltage of 95 volts. With 76 Ni-H cells and 60 plates/cell in series this produces 95V. In January 2017 NASA replace Ni-H with Li-ion batteries These are twice as powerful, so only 3 batteries per side are needed. JAXA delivered these batteries on their HTV vehicle.

Figure 12.8 The HTV ground team installs the cargo pallet holding the space station's six new lithium-batteries inside the supply ship before launch. Credit: JAXA

Cooling Systems

The SAW have radiators, and so does the rest of the station. Looking at figure 12.4 and 12.3 you see that the P1 and S1 trusses have three heat rejection radiators. These trusses have ammonia pumps that flow 290 kg (637 lb) of anhydrous ammonia through the radiators. The rest of the Space Station has heat exchangers, one side has water flowing in it, the other has the ammonia. The heat flows from the water to the ammonia, the ammonia carries the heat to the radiators. The heat exchangers are external to the habitable areas of the Space Station for safety reasons. Why use ammonia? It is not because of the ozone depletion properties of freon. Ammonia is a very efficient coolant, and is used for commercial ice production facilities. In the early part of the 20th century, household refrigerators used it. It was banned because it is toxic in enclosed spaces, such as a kitchen.

During periods of Station shadow, the radiators are oriented towards the Earth to absorb heat from the dark side of the Earth, which still radiates in infrared, so that the Station would not be too cold.

Other Power Systems Planned (but never used)

Solar Dynamic (SD)

To achieve higher efficiencies and power while using less mass and volume SD systems have been tested. Large volumes of solar arrays, such as on the International Space Station, have a large "sail area", which means that they behave as a sail on a sailing ship. The particles in Low Earth Orbit (LEO) hit the arrays and slow the Station down. Although there are not many particles in LEO, the Station is moving very quickly, and the sail area effect causes the constant need to re-boost the Station back to a higher orbit. SD systems reduce this need tremendously. There have been studies to fly Solar Brayton power generators on the International Space Station. The original Space Station Freedom (Figure 12.2) was going to replace the Solar Array Wings (SAW) with Solar Brayton power generators. A joint Russian-US SD system for Mir was cancelled in 1996 after several successful tests. The principle is that a parabolic, or nearly parabolic reflector would concentrate solar energy on a receiver. The receiver has a thermal energy storage (TES a $LiF-CaF_2$ phase change material) unit and a fluid that generates the power in a Brayton cycle. The TES supplies heat when the Station is in Earth's shadow. The company Allied Signal built the flight hardware for the Mir Program and it is currently stored at the Glenn Research Center. The Glenn Research Center tested an SD in their thermal vacuum facility in 1995, it is not clear whether this was the same hardware. During the test the SD produced 2 Kwe. The turbines operated at 52,000 rpm with turbine inlet temperature of 1063.5 K (790.5 C) and a compressor inlet temperature of 270.2 K (-2.8 C). Overall system efficiency was 17% (ratio of electrical energy output to solar energy collected over an orbit), engine cycle efficiency was 29% (ratio of alternator power to heat input). NASA planned to use most of the Mir hardware and redesign the system and install it on the Station, but this was cancelled.

Figure 12.9 Solar Brayton testing at Glenn Research Center (Courtesy NASA)

Figure 12.10 Brayton power conversion unit on left, heat receiver on right. (Courtesy NASA)

Flywheel Energy Storage (FES)

As we mentioned in Chapter 1, NASA was developing a Flywheel Energy Storage (FES) system for the Space Station. When the Station enters the shadow of the Earth, the batteries supply the power to the station through the Battery Charge/Discharge Unit (BCDU). The FES would replace the batteries and the BCDU. The flywheels would store more energy than the batteries, thus allowing extra electrical capacity for the Station and its experiments. Batteries have a limited amount of charge/discharge cycles. Flywheels can be charged/discharged virtually forever. Therefore they last longer and should save 100's of millions of dollars over the life of the Station. Testing at JSC would have started in 2001 with a planned launch to the Space Station in 2004. The project was called the Attitude Control and Energy Storage Experiment (ACESE). It would have been launched on flight 13A. The goals were to demonstrate operation as an energy storage system and to generate differential torques, which can be used to change spacecraft orientation (it is not clear whether it would change the orientation of the Station, or merely generate torques). There would be two counter rotating flywheel modules and controls inside a standard Orbital Replacement Unit (ORU) box with 2.4 Kw-hrs of energy storage. The flywheel rotors will be of advanced composites spinning at 60,000 rpm levitated on magnetic bearings to minimize loss.

Chapter 13, The Future: Fusion, Cold Fusion, and Antimatter

Fusion

There is a lot of world wide research going on in fusion power. Its advantages over nuclear power are that there is virtually an unlimited supply of hydrogen for the reaction from our oceans and it does not have the shielding problems or waste problems that reactors have. For space applications these latter two are important. Putting shielding into orbit and getting rid of the waste are expensive propositions. The power output of fusion is virtually unlimited, ideal for very long and high power missions. The problem with fusion for space power is the weight of the fusion power plant.

Fusion is the opposite of fission. In this case we shove two particles together, forming a new atom. This occurs at a very high energy (i.e. a temperature of hundreds of millions of degrees K). There is a release of energy and this can generate power.

There are two main paths to fusion. One is the magnetic confinement approach and the other is inertial confinement, the latter of which appears to be the most promising in the near term. There is also a third path to fusion called Inertial-Electrostatic Confinement.

The equation of interest is a Deuterium Tritium (D-T) reaction. Deuterium is a Hydrogen isotope with a proton and one neutron, Tritium is a Hydrogen isotope a proton and 2 neutrons. The reaction is:

$$_1^2D + {}_1^3T \rightarrow {}_2^4He^{++} \text{ (3.5 MeV)} + {}_0^1n \text{ (14.1 MeV)} \qquad \text{13-1}$$

The Helium is a charged particle. The high energy neutron is slowed in a blanket surrounding the plasma. The plasma contains the D and T. Depending on the design, the charged Helium's kinetic energy can be used either in a direct energy conversion device, for heating the plasma, or for blanket wall heating (figure 13.1).

Because tritium does not occur naturally, it is necessary in the D-T reaction to breed tritium in the blanket. The lithium reactions that occur in the blanket are:

$$_0^1n + {}_3^6Li \rightarrow {}_1^3T + {}_2^4He \text{ (4.6 MeV)} \qquad \text{13-2}$$

$$_0^1n + {}_3^7Li \rightarrow {}_0^1n + {}_1^3T + {}_2^4He \text{ (-2.4 MeV)} \qquad \text{13-3}$$

The blanket contains the lithium and must integrate three principal functions of a D-T system:

(1) Must regenerate tritium

(2) Must convert the neutron's KE to thermal energy and then make this energy available to a power plant

(3) The structure must serve as an adequate enclosure for the heat transfer fluid. This last is the most difficult.

If the particles in the plasma are allowed to bounce off the containment vessel, two things would happen. The plasma particles will be cooled immediately and quench the reaction and the particle impact on the walls will sputter off wall material, which will increase plasma radiation loss and further accelerate cooling. This problem of quenching led to the first concept, magnetic confinement.

Magnetic confinement

If we confine the plasma by using the lines of force of a magnetic field to hold the hot, ionized plasma in a "magnetic bottle" then the plasma never touches the walls. Many of the fusion devices being developed use this method, particularly the Russian "Tokamak" machines. The problems thus far have been to achieve plasma temperatures of 10^8 K, particle density of 10^{15} ions/cm^3, and confinement times of a few tenths of a second. If the density can be increased, the times can decrease. The break even point (the point at which we get as much energy out as we put into the magnets, controlling devices, etc.) has yet to be reached.

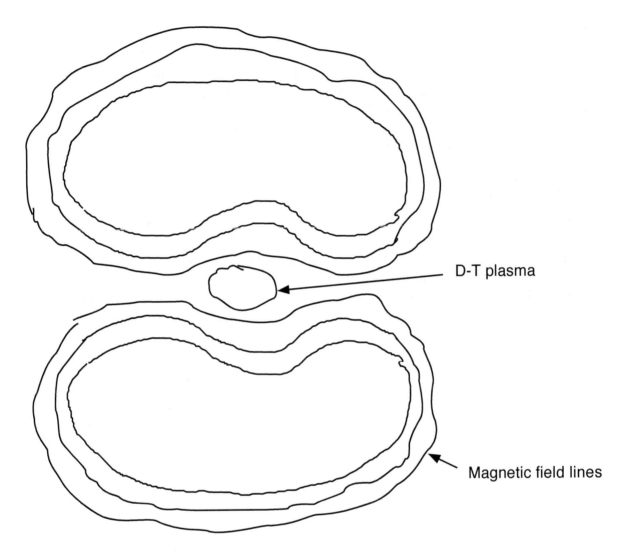

D-T plasma

Magnetic field lines

Figure 13.1 magnetic confinement bottle

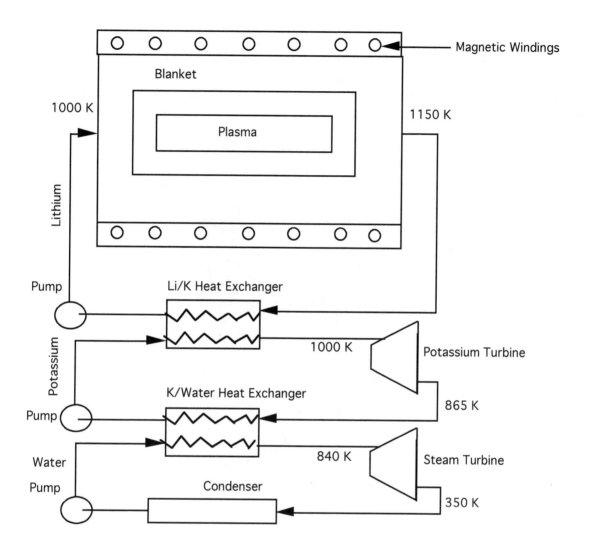

Figure 13.2 Fusion Power Plant

Consider the ITER (pronounced as in "fitter") power plant, the newest fusion research plant being built in Europe. It will not produce power above what it consumes. It is based around a hydrogen plasma torus operating at over 100 million degrees C, and will produce 500 Mw of fusion power. It will not be completed until 2035, although experiments will start in 2025.

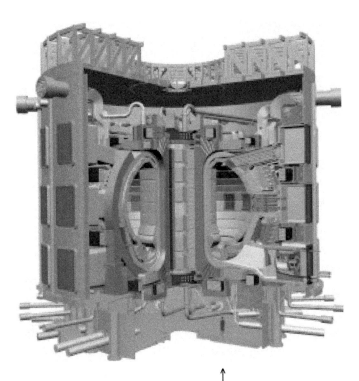

↑

Notice the man (blue) standing at the ITER

Figure 13.3 ITER

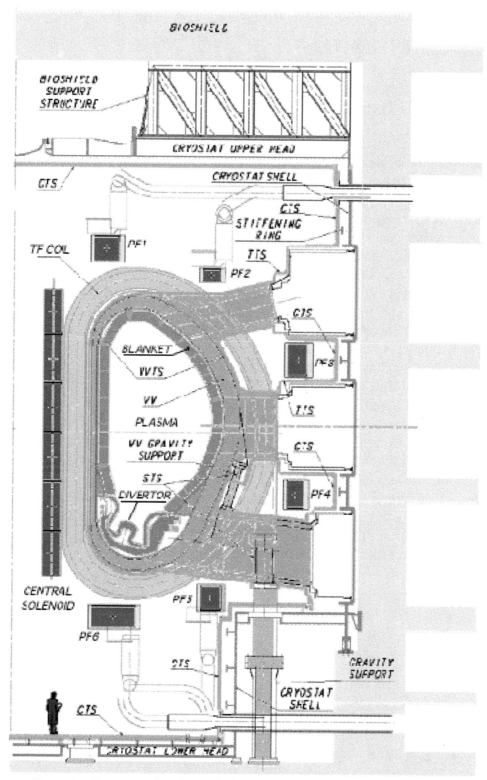

Graphics Courtesy the ITER web site http://www.iter.org/index.htm

Figure 13.4 ITER fusion plant

Inertial confinement

This approach uses intense laser or particle beams to heat and compress "pellets" of Deuterium-Tritium (D-T) until they fuse. There are two methods being proposed to accomplish this.

In the direct drive method, hydrogen contained in a tiny glass sphere is directly irradiated by multiple laser beams. The ablation of the glass sphere causes the generation of imploding shock waves that compress and heat the hydrogen. In the indirect drive method, the laser beams are first transformed into X-rays and then irradiate the pellet.

Recently it has been suggested that combining inertial and magnetic confinement could produce better results, and experiments by the Japanese support this. This method would use less powerful magnetic fields.

At the Lawrence Livermore National Laboratory they are researching indirect drive fusion. This program is happening at the National Ignition Facility. They are using a 2mm spherical plastic container filled with 150 picograms of D-T and supercooled to 18K, where it freezes. It is then put into a container called a hohlraum, which is German for "hollow space". The hohlraum is a cylinder 5 mm in diameter and 9 mm high. It has a seven-micron-thick gold and uranium-layer, the whole thing encapsulated in gold to prevent oxidation and provide structural rigidity. Finally, 192 lasers that can, for 20 billionths of a second generate 500 billion watts, are focused on the ends of the hohlraum. The gold produces a super hot plasma which radiates soft x-rays uniformly on the fuel pellet. The surface of the pellet ablates, or blows off, extremely quickly, imploding the pellet. Isaac Newton's third law states that for every reaction there is an equal and opposite reaction. Therefore the remaining part of the pellet is driven inwards in an implosion. This compresses the fuel and causes a shock wave, which further heats the fuel in the center. The D-T fuel is compressed to a density 100 times that of solid lead, and heated to more than 100 million degrees Celsius – hotter than the center of the sun. Thus, Fusion is achieved. This fusion should produce 100 times the energy required to make the fusion occur.

Figure 13.5 Radiograph of high density carbon capsule with frozen D-T inside, courtesy Lawrence Livermore National Security, LLC, and Lawrence Livermore National Laboratory

Figure 13.6 **Left:** A 2-mm-diameter polished beryllium capsule with a 10-micron fill tube attached. **Right:** 2-mm polished high-density carbon ablator capsules with the silicon mandrel inside. Photos courtesy Lawrence Livermore National Security, LLC, and Lawrence Livermore National Laboratory

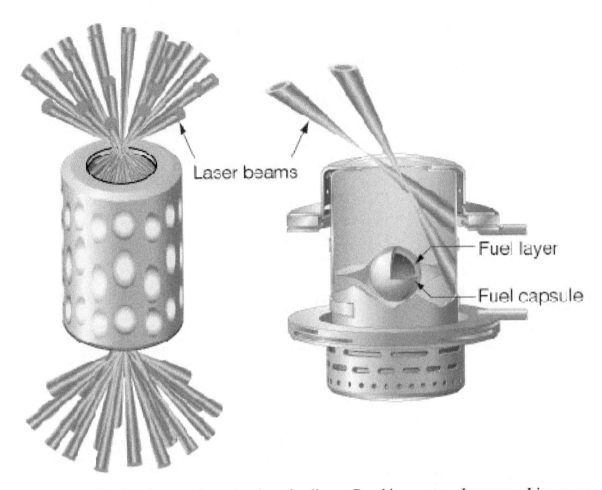

Figure 13.7 The Hohlraum, about the size of a dime. Graphic courtesy Lawrence Livermore National Security, LLC, and Lawrence Livermore National Laboratory

Inertial-Electrostatic Confinement Fusion

There has also been some work on fusion using electrostatic acceleration instead of temperature. These systems are generally called Inertial-Electrostatic Confinement (IEC). Research has been going on for several years at University of Wisconsin-Madison, and they have achieved fusion at small levels. The basic device is a vacuum chamber with a fuel such as deuterium. It uses an inner spherical grid of Tungsten that is is 90-99% transparent and is charged to -100 kV. The outer spherical grid is at nearly zero potential. The potential difference between the grids accelerate the ions to fusion velocities.

Figure 13.8 the IEC 10 cm diameter grid in operation. Photo courtesy the University of Wisconsin-Madison

Figure 13.9 IEC chamber. Photo courtesy the University of Wisconsin-Madison

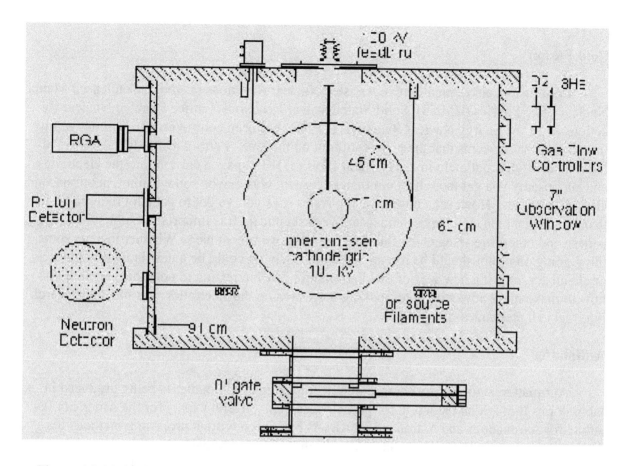

Figure 13.10 the IEC in operation. Photo courtesy the University of Wisconsin-Madison

Positive ions are created at the outer grid, and are accelerated towards the negatively charged inner grid. The ions may shoot past the inner grid, then turn around and go back towards it. This oscillation creates a concentration of high temperature ions. The ions can collide, creating a fusion reaction. The ions can also grab an electron, creating a fast neutral particle. The fast neutrals can collide with the neutral gas, creating fusion reactions. Depending on the fuel,

Some of the reactions that the University has investigated are:

$$_1^2D + _2^3He \rightarrow _2^4He \ (3.67 \text{ MeV}) + _1^1p(14.68\,MeV)$$

$$_1^2D + _1^2D \xrightarrow{50\%} _1^3T \ (1.01 \text{ MeV}) + _1^1p(3.02\,MeV) \qquad 13\text{-}4$$

$$_1^2D + _1^2D \xrightarrow{50\%} _2^3He \ (0.82 \text{ MeV}) + _0^1n(2.45\,MeV)$$

This is promising research, but many years away from powering spacecraft.

Cold Fusion

Cold fusion, also called Lattice Assisted Nuclear Reactions (LANR), Condensed Matter Nuclear State Physics (CMNS), Solid State Nuclear Reactions, Lattice Enabled Nuclear Reactions and Low Energy Nuclear Reactions (LENR) has been controversial. Without getting into the history, it seems that the jury is still out on this one. Putting it simply, take two Palladium electrodes, put them in heavy water (D_2O vs. H_2O), pass a current into the electrodes and supposedly you get more heat out than energy in, plus maybe some tritium, neutrons, and charged particles. However, conventional physics says you've got to get a lot more neutrons than have been seen, so many scientists are very skeptical. It is important to note that the Japanese and others are researching this, and the US government isn't. Whether there is something going on or not should be researched extensively. It could be a new branch of physics, or chemistry, or just a new way of making money. It is interesting to note that research into how to make more power involves making more area on the electrodes, just like the research into fuel cell electrodes.

Antimatter

Antimatter is not just for science fiction spaceships. Antimatter is being produced in laboratories throughout the world for particle research. Several papers for the American Institute for Aeronautics and Astronautics (AIAA) have been written proposing methods for producing and storing significant quantities of anti-protons. Anti-Protons are created by colliding beams of high energy protons with heavy element targets. This produces a lot of secondary particles, some of which are relativistic anti-protons. These are separated from the other particles, cooled, and placed in a storage ring. To be useful as a power source, they must be combined with anti-electrons to produce neutral anti-hydrogen. The anti-hydrogen must be stored to keep it from touching matter. This means that it must be cooled to become anti-hydrogen ice and then magnetically suspended at temperatures near absolute zero so that no reactions occur due to the vapor pressure of both the anti-hydrogen and the container. Once stored, the anti-hydrogen ice is reacted with hydrogen to produce hot charged particles, gamma rays, and neutrinos. The latter two are not useful for power production but the charged particles can be used to produce power either by heat transfer, MHD, or some other method. Most of the literature is focused on using this for propulsion, here we will look at it for power production.

Producing Antimatter

Dr. Robert L. Forward (physicist and science fiction author) proposed making antimatter factories using the methods of the current production facilities at IHEP (Russia), CERN (Switzerland) and Fermi Lab (USA) specially designed to maximize the anti-proton production and electrical power efficiencies. Figure 13.11 shows his proposed factory. It consists of several proton beams each operated at optimum beam current. Each beam strikes a thin liquid metal target and the resulting particles would be sorted by an array of magnetic lenses to extract the anti-protons and anti-electrons. The main beam would then go on to a beam cooler to reduce the spreading of the beam and then go on to the next target. The anti-electrons with the correct energy would be collected an sent on to the anti-hydrogen generator. The anti-protons would be sorted by energy and sent to a stack of stochastic coolers, each cooler optimized for a particular anti-proton momentum. After stochastic cooling, the stack of anti-protons at different energies would go to a decelerator stack where all the anti-proton energies would be reduced to the same sub-relativistic energy of 200 MeV. The combined beams would be sent to an electron cooling ring and then sent on to the anti-hydrogen generator (Figure 13.12). If the beams of anti-protons and anti-electrons are traveling at the same speed they will combine naturally and form anti-hydrogen. This process can be made much more efficient (by a factor of 100 or so) by using lasers to stimulate the capture process. The anti-hydrogen is then cooled until anti-hydrogen molecules are formed in a para-hydrogen state . Further cooling using lasers operating on a molecular hydrogen line will turn the anti-hydrogen molecules into ice in the para-hydrogen state.

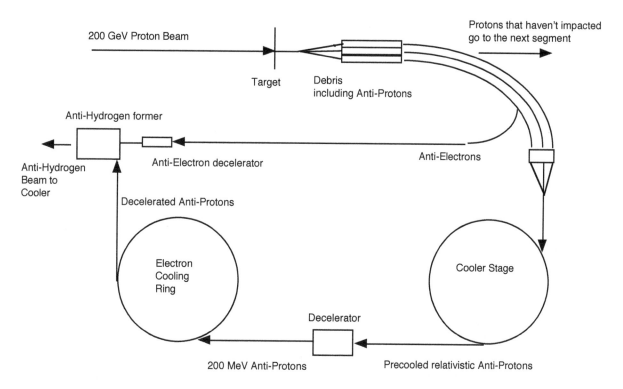

Figure 13.11 One segment of an Anti-hydrogen Factory

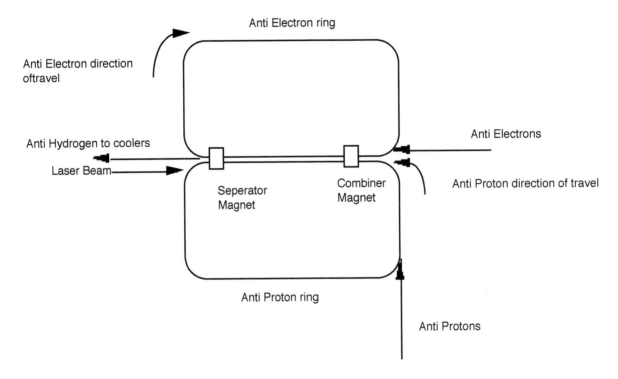

Note: All the rings have magnets to direct the particles, and everything is in a vacuum

Figure 13.12 Anti-hydrogen formed by laser

Storage

Once the ice is produced it must be stored safely. Fortunately the ice is diamagnetic, so a simple magnetic bottle could be made using superconducting magnets. This has the advantage that it is completely stable and requires no power. It has the disadvantage that it could not levitate the ice at high acceleration levels. For high acceleration a servo controlled DC voltage electrostatic levitation (EL) device could be used. EL of large spheres with the density of anti-hydrogen ice (0.0763 gm/cc) has been demonstrated at JPL. The only problem would be the heat input from the levitator sublimating the ice. Fortunately, if the ice is kept cold enough, this is not a problem. The estimate is that at 4 K a milligram of anti-hydrogen ice should last more than 10 years. To accommodate both high acceleration and weightlessness, both methods could be combined. Figures 13.13 and 13.14 show the two storage methods.

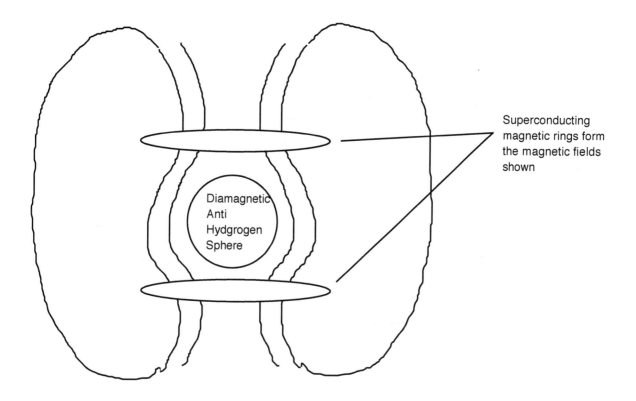

Superconducting
magnetic rings form
the magnetic fields
shown

Diamagnetic
Anti
Hydgrogen
Sphere

The diamagnetic antihydrogen is stabilized by
being levitated in the magnetic field minimum.

Figure 13.13 Levitation of an anti-hydrogen sphere in a stable magnetic field

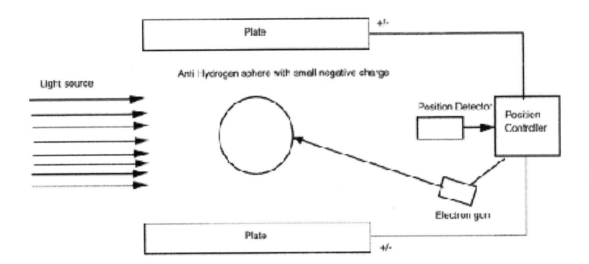

Figure 13.14 Electrostatic levitation

Matter-Antimatter Reaction

The anti-protons are extracted from the ice by irradiating the ice with ultraviolet, driving off the anti-electrons, and then extracting the anti-protons by field emission with a high intensity electric field. They are then directed to a reaction chamber where they combine with matter. The matter can be anything, but hydrogen is chosen for a number of reasons, not the least of which is neutron damage to the device if an atom other than hydrogen is used. When an atom of hydrogen reacts with an anti-proton the following reaction occurs.

$$p + \bar{p} \longrightarrow 1.6\pi^0 + 1.6\pi^+ + 1.6\pi^- \qquad\qquad 13\text{-}5$$

Where p and $\bar{p}$ are protons and anti-protons and π are pions. The charged pions further decay with a half life of 26 nanoseconds into muons (μ). However, since they are moving at 94% of the speed of light, their lives are lengthened to 70 nanoseconds. This means they travel 21 meters before decaying into muons as follows.

$$\pi^0 \longrightarrow \gamma + \gamma$$
$$\pi^+ \longrightarrow \mu^+ + \nu_\mu$$
$$\pi^- \longrightarrow \mu^- + \bar{\nu}_\mu \qquad\qquad 13\text{-}6$$

The neutral pions decay with a half life of 0.84×10^{-16} seconds (10^{-7} nanoseconds!). The neutrinos do not react with matter and the gamma rays are too energetic to react with matter so the energy in these two is lost. The muons have a half life of 2.2×10^{-6} seconds before decaying to;

$$\mu^+ \longrightarrow e^+ + \bar{\nu}_\mu + \nu_e$$

$$\mu^- \longrightarrow e^- + \nu_\mu + \bar{\nu}_e$$

13-7

The charged pions and muons will react with matter and ionize atoms. The neutral pions only react with nuclei and due to their extremely short half life any interaction that we want to use to extract their energy must occur at the reaction site.

Power Generation

The charged particles can be directed by magnetic fields, figure 13.15.

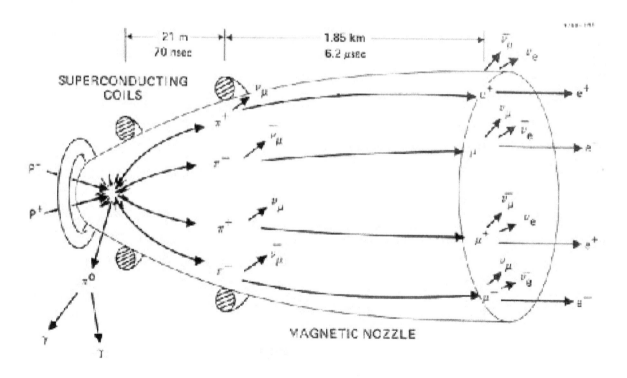

Figure 13.15 from Robert L. Forward.
This was for a rocket nozzle, but the same principles apply

There are a couple of ways to produce power using the reaction products of the matter-anti-matter reaction. If we focus the particles in a beam and make the beam long enough, what we get at the end are electrons and anti-electrons. These can be separated and used in a

MHD generator. Some of the papers want to use the antimatter for propulsion, and they find it more efficient to use the reaction to heat more hydrogen to be used as a propellant. We could do this as well and use the resulting hot gas for a MHD generator or to produce power in a Brayton generator or a thermoelectric generator or for any power source that uses heat or charged particles. B. N. Cassenti in his paper on antimatter for Orbital Transfer Vehicle (OTV) propulsion calculated combustion chamber temperatures for various propellants versus the molecular weight of the propellant. The higher the molecular weight, the higher the combustion temperature. For a generator similar temperatures can be assumed. At the low end, hydrogen could be used to make temperatures of 450 K for use in a thermoelectric generator, whereas water could be used for a 3000 K MHD generator. Cassenti figured an efficiency of 0.35 for conversion of the annihilation energy transferred to the propellant. For a generator, let us assume 0.25 for the efficiency. How much power is this? Today the laboratories can produce 10^{13} anti-protons, which have a mass of 17 picograms. When this is annihilated with an equivalent amount of matter it releases approximately 3056 Joules of energy. Cassenti's rocket was designed to consume about 6×10^{-3} mg/s which is 6×10^6 picogram/s or 1.058×10^9 Watts. Using this with an efficiency of 0.25 means the generator would get 2.64×10^8 W or 264 Megawatts. The power density for the anti-hydrogen is 4.4×10^{10} W/gm, a truly impressive amount. If we take into account the hydrogen that would be used in the reaction we get half that amount, 2.2×10^{10} W/gm, but still impressive. The assumption is that the rest of the hydrogen that is heated is recycled and not lost. Also, the energy used for the magnetic fields, pumps, and in cooling the hydrogen to be recycled back into the reaction chamber is not taken into account. This doesn't matter, if we take all those losses into account and change the efficiency by a factor of 10 we still get an extremely advantageous high power density power source. This is very desirable for space applications.

Conclusion

The use of antimatter is merely a technical and cost issue and well within the realm of possibility. It's advantage of high power density makes it ideal for such missions as a Mars settlement. deep space probes, or remote bases. The generator considered would produce 264 MW for one year and consume only 1890 grams of antimatter. Whether producing and transporting this much antimatter to Mars is cheaper than building an equally powerful fission reactor needs to be considered.

Producing such quantities of antimatter require a high vacuum and real estate that could be "written off" if there was an accident. New Jersey would not be a possibility, but the Moon would be an excellent choice for an antimatter production facility.

References

https://lasers.llnl.gov/ Lawrence Livermore National Laboratories National Ignition Facility and Photon Science Homepage

Forward, R. L.; "Anti-Proton Annihilation Propulsion", AIAA/SAE/ASME 20th Joint Propulsion Conference, June 1984

Cassenti, B. N. ; "Optimization of Relativistic Antimatter Rockets" AIAA Paper No. 83-1343, 19th Joint Propulsion Conference, 1983

Cassenti, B. N.; "Antimatter Propulsion for OTV Applications", AIAA/SAE/ASME 20th Joint Propulsion Conference, June 1984

Appendix A Units of Measurement

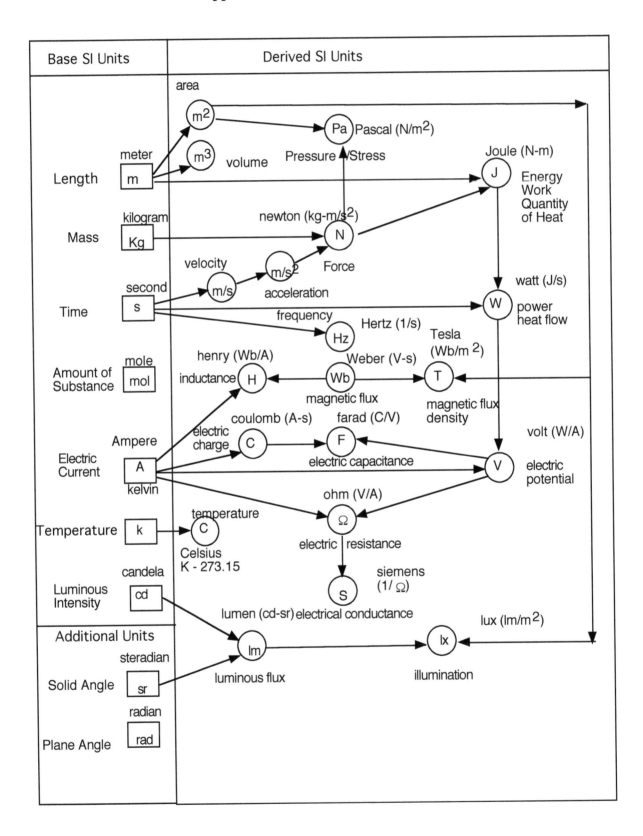

Appendix B Constants and Conversion Factors

R Universal gas constant

8.3144598 J/K-mole
1.9859 kcal/K-mole
0.082057 liter-atm/K-mole

N Avagrado's number

6.022140×10^{23} /mole

k Boltzmann's constant

$1.38064852 \times 10^{-23}$ J/K
$1.38064852 \times 10^{-16}$ erg/K
8.6173303×10^{-5} ev/K

sigma Stefan-Boltzmann's constant

5.670367×10^{-8} W/m^2 -K^4

e electronic charge

$1.6021765 \times 10^{-19}$ C

m rest mass of the electron

9.109383×10^{-31} kg

h Planck's constant

$6.62607004 \times 10^{-34}$ J-s

c speed of light

2.99792458×10^{8} m/s

F Faraday's constant

9.6485332×10^{4} C /mole
23.061 kcal/volt-mole

E_0 permittivity of free space

8.854187×10^{-12} F/m

1 standard atmosphere

760 mm Hg
14.696 lbf/in^2
1.013×10^{5} N/m^2

g standard gravitation acceleration

9.80665 m/s^2

Conversion Factors

1 C (Coulomb)	= 1 amp/second
1 electron volt	= 1.602×10^{-19} joules
	= 1.602×10^{-12} ergs
1 farad	= 1 coulomb/volt
1 joule	= 10^7 ergs
	= 1 newton-meter
	= 2.389×10^{-4} kcal
1 meter	= 10^{-2} cm
	= 10^{-6} microns
	= 10^{10} angstroms
1 newton	= 10^5 dynes
	= 1 kg-m/second^2
	= 1 volt-coulomb/meter
1 ohm	= 1 volt/amp
1 volt	= 1 Joule/coulomb
1 watt	= 10^7 ergs/second
	= 1 joule/second
1 weber	= volt/second
	= 1 joule/amp
1 weber/m^2	= 10^4 gauss
	= 1 newton/amp-meter
	= 1 tesla

Appendix C The Chemical Elements

Atomic Number	Name	Symbol	Atomic Weight
1	Hydrogen	H	1.008
2	Helium	He	4.003
3	Lithium	Li	6.941
4	Beryllium	Be	9.012
5	Boron	B	10.81
6	Carbon	C	12.01
7	Nitrogen	N	14.01
8	Oxygen	O	16
9	Florine	F	19
10	Neon	Ne	20.18
11	Sodium	Na	22.99
12	Magnesium	Mg	24.31
13	Aluminum	Al	26.98
14	Silicon	Si	28.09
15	Phosphorus	P	30.97
16	Sulfur	S	32.07
17	Chlorine	Cl	35.45
18	Argon	Ar	39.95
19	Potassium	K	39.09
20	Calcium	Ca	40.08
21	Scandium	Sc	44.96

Atomic Number	Name	Symbol	Atomic Weight
22	Titanium	Ti	47.88
23	Vanadium	V	50.94
24	Chromium	Cr	52
25	Manganese	Mn	54.94
26	Iron	Fe	55.85
27	Cobalt	Co	58.93
28	Nickel	Ni	58.69
29	Copper	Cu	63.55
30	Zinc	Zn	65.39
31	Gallium	Ga	69.72
32	Germanium	Ge	72.58
33	Arsenic	As	74.92
34	Selenium	Se	78.96
35	Bromine	Br	79.9
36	Krypton	Kr	83.8
37	Rubidium	Rb	85.47
38	Strontium	Sr	87.62
39	Yitrium	Y	88.91
40	Zirconium	Zr	91.22
41	Niobium	Nb	92.91
42	Molybdenum	Mo	95.94
43	Technetium	Tc	98
44	Ruthenium	Ru	101.1

Atomic Number	Name	Symbol	Atomic Weight
45	Rhodium	Rh	102.9
46	Palladium	Pd	106.4
47	Silver	Ag	107.9
48	Cadmium	Cd	112.4
49	Indium	In	114.8
50	Tin	Sn	118.7
51	Antimony	Sb	121.8
52	Tellurium	Te	127.6
53	Iodine	I	126.9
54	Xenon	Xe	131.3
55	Cesium	Cs	132.9
56	Barium	Ba	137.3
57	Lanthanum	La	138.9
58	Cerium	Ce	140.1
59	Praseodymium	Pr	140.9
60	Neodymium	Nd	144.2
61	Promethium	Pm	147
62	Samarium	Sm	150.4
63	Europium	Eu	152
64	Gadolinium	Gd	157.3
65	Terbium	Tb	158.9
66	Dysprosium	Dy	162.5
67	Holmium	Ho	164.9

Atomic Number	Name	Symbol	Atomic Weight
68	Erbium	Er	167.3
69	Thulium	Tm	168.9
70	Yttrium	Yb	173
71	Lutetium	Lu	175
72	Hafnium	Hf	178.5
73	Tantalum	Ta	180.9
74	Tungsten	W	183.9
75	Rhenium	Re	186.2
76	Osmium	Os	190.2
77	Iridium	Ir	192.2
78	Platinum	Pt	195.1
79	Gold	Au	197
80	Mercury	Hg	200.5
81	Thallium	Tl	204.4
82	Lead	Pb	207.2
83	Bismuth	Bi	208.9
84	Polonium	Po	209
85	Astatine	At	210
86	Radon	Rn	222
87	Francium	Fr	223
88	Radium	Ra	226
89	Actinium	Ac	227
90	Thorium	Th	232

Atomic Number	Name	Symbol	Atomic Weight
91	Protactinium	Pa	231
92	Uranium	U	238
93	Neptunium	Np	237
94	Plutonium	Pu	242
95	Americium	Am	243
96	Curium	Cm	247
97	Berkelium	Bk	247
98	Californium	Cf	249
99	Einsteinium	Es	254
100	Fermium	Fm	253
101	Mendelevium	Md	256
102	Nobelium	No	254
103	Lawrencium	Lr	257
104	Rutherfordium	Rf	257
105	Dubnium	Db	260
106	Scaborgium	Sg	263
107	Bohrium	Bh	262
108	Hassium	Hs	265
109	Meitnerium	Mt	266
110	Darmstadium	Ds	271

References

Angrist, Stanley W.; "Direct Energy Conversion, Fourth Edition", Allyn and Bacon, Inc., 470 Atlantic Avenue, Boston, MA 02210, 1982 ISBN 0-205-07758-7

Angelo, Joseph A. and Buden, David; "Space Nuclear Power". Orbit Book Company, Inc. 2005 Township Road, Malabar, FL 32950, 1985 ISBN 0-89464-000-3

Printed in the USA
CPSIA information can be obtained
at www.ICGtesting.com
LVHW072022110124
768644LV00016B/563